WILD HARVEST

An Outdoorsman's Guide to Edible Wild Plants in North America

WILD HARVEST

ALYSON HART KNAP

Illustrated by E. B. Sanders

Pagurian Press Limited

Distributed by
Publishers Marketing Group
A Baker & Taylor Company
Executive Offices
1515 Broadway, New York, N.Y. 10036

Distribution Center
Gladiola Avenue, Momence, Ill. 60954
ISBN 0-919364-97-7

In memory of

Luigi and Marthe Colombani,

who first introduced me to the wild harvest.

TABLE OF CONTENTS

PREFACE

The out-of-doors has had a tremendous upswing in popularity in recent years. Many more people today are taking to the fields and forests than did a generation or even a decade ago. This surge away from the confines of the city to the freedom of nature is evidenced by the increased numbers of people in our parks and campgrounds, on our lakes and rivers, and in our wilderness areas. It is also attested to by the rising sales volume of outdoor equipment in the recent past. And lastly, it can be seen in the number of books on wildlife, nature lore, and outdoor skills that have been published over the past few years. Among these books are a number of volumes dealing with the edible wild.

The literary pieces that have appeared on wild edibles are generally of two types: field guides describing the edible wild plants, with their habitats and life styles; and cookbooks dealing with recipes for preparing, cooking, and preserving wild fruits and herbs after the harvest. In this book, an attempt has been made to tie both together; to take the wild harvester into the field for his harvest and to guide him in the preparation and enjoyment of his gleanings.

There are a great number of edible wild plants on this continent, enough to fill several volumes of this size with descriptions and kitchen hints. Hence the number of plants described in this book has had to be somewhat restricted. The species covered are mostly popular or ubiquitous ones; plants that can easily be recognized by the tyro, and long-time favorites of the "old hands."

Although the plants have been categorized according to what we feel are their better culinary uses, it should be realized that with a few stated exceptions, the only limits in readying the wild edibles for the table lie in the imagination and culinary talents of the chef in charge of their preparation. Most edible wild plants have amazing diversity and adaptability in the kitchen.

The wild harvester needs surprisingly little equipment to enjoy his outdoor pastime. Often a small knife and a basket are all that is necessary. Knowledge of the plant kingdom and its members comes quickly with exposure and experience, but the wild harvester may find a botanical field guide, describing in detail the many species that grow in specific areas, to be of help.

There is no limit to the enjoyment that can be had from a venture away from the bustle and worries of civilization, into the refreshing solitude of the out-of-doors. A wild harvest can serve to soothe and calm the nerves and soul, and can also provide new and interesting taste experiences.

With this book come best wishes for many enjoyable and memorable wild harvests.

INTRODUCTION

We are all aware of the plight of the sea otter, the bison, and the fur seal. They are but a few examples of wildlife resources that came near extinction at the hands of greedy and wasteful men. Yet one rarely hears about plants that have gone the same route. Here is the story of one such plant, the herb *Panax quinquefolium* — the North American ginseng.

In the year 1727, a Jesuit missionary by the name of Father Lafitau was the parish priest to the Catholic Iroquois of Caughnawaga, across the St. Lawrence River from Montreal. Like many clerics of his time, Father Lafitau was deeply interested in nature. He observed the Caughnawagas gathering the roots of a plant to which they attributed great medicinal properties. The roots were dug only after the plant's red berries appeared in the fall, and the berries were always then buried in the soil. The Indians wisely claimed that this practice ensured future crops. They called the plant *garent-oguen,* a name derived from *orenta oguen* meaning "legs of man."

All medicinal herbs were of great interest during this period, thus Father Lafitau examined the plant in great detail. The herb strongly resembled a medicinal plant of the Chinese which had been described by Father Jartoux in the *Philosophical Transaction of the Royal Society of London* in 1714. The similarity between these two plants extended even to their names, as the Chinese name meant "man plant" or "man's legs."

Father Jartoux not only described the plant, but also wrote of its great commercial value in China. The demand by the Chinese for this plant greatly exceeded the supply. On the basis of this, Father Lafitau sent a few samples of the North American ginseng to Paris and on to Peking. It took nearly two years before he received word that Peking merchants would purchase all the roots that could be gathered. The

Caughnawagas shared the news with their Indian brothers further south. Within a short time, even the white colonists were gathering the roots. Thus the great ginseng hunt began. Every woodlot and forested hillside was searched. At first ginseng was very common, but by the turn of the century the plants were scarce in all but the most inaccessible wildwoods. Within another three- to four-score years, the wild ginseng hunt was almost over. The plant faced near extinction.

The trading fleets and caravans of the Chinese merchant princes plied their trade with Korea and Manchuria. Their cargoes of silk, pottery, and tea were traded for bundles of dried roots of the Asiatic ginseng — *Panax ginseng*. These roots had been a commodity of great value in the commerce of China long before Marco Polo visited Kubla Khan's Celestial Kingdom. The ginseng roots were boiled in silver kettles and the beverage obtained was reputed to have great therapeutic powers.

Sir Edwin Arnold, the oriental scholar, gives us some idea of the properties that the Chinese attributed to the plant in his book, *The Light of Asia*. "According to the Chinamen, ginseng is the best and most potent of cordials, of stimulants, of tonics, of stomachics, cardiacs, febrifuges, and, above all, will best renovate and reinvigorate failing forces. It fills the heart with hilarity, while its occasional use will, it is said, add a decade of years to the ordinary human life."

The value of the early North American ginseng trade is unknown. At first the Company of the Indies, which held France's trading monopoly with the Orient, allowed free trade. But in 1751 the company attempted to take over the trade on its own behalf. The profits were too great to be shared. This monopoly meant a lower price for the roots to the Indians and coureurs de bois, who immediately transferred their business to the British companies further south.

The first truly big cargo of ginseng was some fifty-five tons on the British sloop *Hingham* in 1773. John Jacob Astor took with him the first direct shipment of ginseng in 1782 on his trip from New York to China. The cargo sold for three dollars per pound. Two years later the American ship, the *Empress of China*, repeated the journey. Even Daniel Boone took part in the ginseng hunt. In 1788 he sold fifteen tons of ginseng root to a Philadelphia export firm.

Just what does this ginseng look like? The plant is a short herb, ranging from eight to fifteen inches in height. It bears three leaves at its apex, each consisting of five leaflets — three large upper ones and two smaller lower ones. The leaflets are oval shaped, pointed at one end, and have toothed margins. During July and August a cluster of small, rather inconspicuous, greenish-yellow flowers is produced, but the plant is much more easily recognized in the fall when its bright crimson berries appear.

The root of the ginseng is thick and spindle-shaped. It varies from two to four inches in length and from one-half to one inch in thickness. It is pale yellow to yellowish-brown in color, fleshy before drying, and has a wrinkled skin. After it has been properly dried, it is firm and solid. During its first year of growth, the root is straight; as the plant matures, the root branches. Occasionally these branches assume the shape of a human body, and such a root is much more valuable to the Chinese than a straight or oddly branched root. The root has a slightly aromatic odor and a sweet, rather pleasant taste.

Ginseng grows in rich, deep soil and requires protection from the direct rays of the sun. This is supplied by the tall forest canopy. It is rarely found near wet ground or stagnant water, although it may grow on the banks of running forest streams. The root is slow in maturing. Five to seven years of growth are required before the roots become marketable.

The range of the wild American ginseng is quite wide. The plant grows in the temperate climate of eastern North America from southern Quebec to the Carolinas, and as far west as the Mississippi River. However, due to persistent overharvesting, it is now scarce over most of its range.

Perhaps it should be stated at this time that western pharmacologists do not consider ginseng to have any medicinal properties whatsoever. The generic name *Panax* is a reminder of the era when the prime concern of botany was medicinal herbs. The name is a derivative of the Greek word *panacea,* meaning "cure-all." The use of ginseng on this continent was limited. It was one of the ingredients in a patent medicine known as "Garfield Tea," and was also used as a flavoring in several brands of chewing tobacco.

The ginseng trade has long been a paradox. A plant of no known medicinal value was brought to nearly total elimination by people seeking it for medicinal purposes.

It is because of the current surge of interest in edible wild plants that I have introduced this text with the near-tragic story of the ginseng. The tenets of resources management dictate that an excess of every crop is provided each year by nature, and that such excess can be harvested.

But an overharvest must not be allowed to take place. A wild harvest must be carried out with temperance and prudence, otherwise our plant resources will suffer, and when that happens we are all losers.

North American Ginseng

Chapter I

SALADS FROM NATURE

It has only been in the past three or four decades that we in North America have come to appreciate fresh greens, eaten without cooking. By and large, our culinary culture before that time dictated that all vegetables be eaten cooked, and in most cases that meant cooked to death. What brought salads onto our menus is anyone's guess. This healthy change was probably caused by a multitude of things, ranging from ease of preparation (no pots or pans to deal with) to the realization that raw greens actually taste quite good.

If you are a fresh salad fan, you can really get hooked on nature's own salad greens — the kind you collect from pastures, roadsides, bogs, dense stands of brush, and many other places, both wild and not so wild. If you are a beginning wild harvester, start with places that are not so wild, like your back lawn. Pick up a mess of young, tender, dandelion leaves, toss them into a salad, and enjoy the highly nutritious and pleasant taste experience. Then, on a Saturday morning, try collecting greens from a nearby pasture or meadow. You will find that wild greens are available to you in amazing abundance and variety from the first hint of spring until long after the local mallards have decided to push on southward. To boot, they are higher in vitamins and minerals than cultivated greens, and after you pick them you don't have to line up at the cash register.

DANDELION

Taraxacum spp.

Many a child's first bouquet of flowers for his mother is an assortment of seed heads of dandelion. And what a heartbreak it is when the treas-

15

ured flowers arrive home almost bare, the seeds having floated merrily away on a breeze. The dandelions are a very prolific species and provide fun for children in gathering and blowing the seed from the inflorescence; tasty salads for edible-wild hunters; robust and tasty ingredients for the wine buff, and excellent emergency rations for the outdoorsman.

Dandelions, particularly the common species, have a long history of use as an edible and a medicinal cure. European settlers in North America brought with them the idea of dandelion tea as a heartburn remedy. Some Indian tribes drank tea made of dandelion leaves for its tonic properties. Kiowa Indian women are even reported to have consumed dandelion flower tea to relieve cramps and dull menstrual pain. (It should be tried!)

Anyone who has ever had a front lawn to care for knows the plant with leaves in the shape of a rosette, milky sap, and bright, sunny yellow, composite flower heads. It seems that no matter how hard the home-turf manager tries to get rid of the little beasts, there they are again! So harvest them and use them for dinner.

There are many species and subspecies of dandelions, all varying slightly in the shape of the leaves and the composition of the flower head. But to anyone who is not a taxonomic botanist (me included), a dandelion is a dandelion. The common dandelion, *Taraxacum officinale,* is a plant of lawns, grasslands, and open areas. It blooms from March to September, depending on the area, and is normally perennial in nature. Dandelions proliferate throughout the cold and temperate regions of this continent.

By far the most common use of dandelion greens is in a salad. Prepared with small chunks of tomatoes, crumbled bacon, small bits of sharp cheddar cheese, and an oil and vinegar dressing, they are a delight. Young leaves are the mildest and most tender; older leaves tend to have a somewhat bitter flavor. Young flower buds, collected before the flowers have come into bloom, also provide good salad makings.

Dandelion roots (young ones) can also be eaten, after peeling off the outer skin. Boiling is the most common way of preparing them, then sprinkling with a shade of salt and some butter. They are quite nutritious, and have saved many a lost outdoorsman and many a settler during times of hunger. Dandelions are very rich in vitamins A and C.

Dandelions can be used for all sorts of beverages — tea, coffee, and wine. Tea is easy to prepare with either leaves or flowers. Coffee requires roasted and ground roots, and the end product is "over 97% caffeine free," and good too. Wine is made from the flowers, and the results can be truly superb.

As a cooked vegetable, young, crisp, and tender dandelion leaves can be boiled for a few minutes with water and served like spinach or kale. My family and I have also had some success in blanching and freezing them. One of our favorite recipes calls for lightly tossing the leaves in a small amount of butter in a non-stick frying pan. Then add the leaves to prepared and previously cooked mushroom slices, chopped onions, and a half clove of minced garlic. Once this is all together in a casserole, add salt and pepper to taste, and enough sour cream to moisten the leaves and turn the whole dish into a creamy delight. Then pop it into the oven for a few minutes while the venison chops are retrieved from the barbecue and the garlic bread is sliced.

LAMB'S-QUARTERS
Chenopodium spp.

Dandelions are not the only green leaves that provide flavor and variety at the dinner table. One of the most common wild plants collected for the pot is lamb's-quarters, a relative of the cultivated spinach. The leaves of the wild variety are not quite as crisp or as large as the farm-grown type, and the taste is not as strong.

Lamb's-quarters grow widely throughout the central and northeastern states, including adjacent areas of Canada; and indeed spread in lesser quantity over a much greater area of this continent. These annual weeds are characterized by a mealy or granular, whitish underside on the leaves. The leaves may not look terribly appetizing, but they are a real deception. The leaves somewhat resemble a goose's foot in shape, hence the genus name *Chenopodium,* from the Greek *chen* meaning goose

and *pous* meaning foot. Small clusters of spike-shaped flowers develop in late summer and early fall, giving rise to small, black, hard seeds.

Plants of lamb's-quarters are commonly seen as weeds in home vegetable gardens in early spring, before planting of domestic species is done. Oddly enough, many a v i d gardeners destroy young wild spinach plants in order to sow cultivated spinach seed! Lamb's-quarters also proliferate along roadsides, in ditches, near construction sites, and in other areas where the soil is frequently worked up or otherwise disturbed.

Pigweed or goosefoot, as lamb's-quarters are sometimes called, was used by many Indian tribes as a foodstuff. The tiny, hard seeds were ground and used in baking bread. They were also added to cornmeal in other baked goods. Even Napoleon Bonaparte is reported to have lived for long periods of time on bread baked from pigweed seeds.

Young plants, less than a foot in height, are generally best for the table. Even when the plants become older, however, the upper parts — the new leaves and the tips — are usually still tender. The young plants can be harvested and used as leafy green vegetables in many of your favorite spinach or Swiss chard dishes. Boil the leaves, drain them, and then add them to a skillet with sautéed small onion or leek. Then add a few drops of wine vinegar, salt and pepper to taste, and a small teaspoon of sugar, and simmer for 10 minutes. Vinegar or lemon juice tend to preserve the high vitamin content of lamb's-quarters, and also add a bit of flavor to the very mild leaves.

Bread made with ground seeds of lamb's-quarters is very dark in color, but has a somewhat insipid flavor. The seeds present quite a challenge to grind, because they tend to slip and slide, and end up all over the kitchen. Some cooks boil them for a few hours, grind up the product, and then dry the mash. This is a bit easier than trying to grind the hard seeds. We have made corn meal bread using corn meal and ground pigweed seeds about fifty-fifty, and the dark brown hue that results makes the bread look very rich indeed. It's a treat for breakfast, smothered with fresh butter and blackberry preserves.

STRAWBERRY BLITE

Chenopodium capitatum

The name "strawberry blite" somehow doesn't convey the idea of a fresh, leafy green salad. Perhaps this plant's other name, strawberry spinach, gets the idea across better. And indeed the strawberry blite is another of nature's wild spinaches, free for the taking.

Strawberry spinach, also known by the third name of Indian paint, is a close relative of lamb's-quarters, and almost as good on the table. The plant is an upright annual, growing to a couple of feet in height. One of the features that distinguishes it from pigweed is its triangular-shaped leaves. With the length of the leaf being greater than the width, a distinctive isosceles triangle appears, with coarsely indented edges. The main difference between the two species of *Chenopodium*, however, is the presence in late summer of masses of soft, strawberry-red fruit on stalks arising from the leaf axils of the strawberry spinach.

Strawberry blite is a plant of light soils. It favors edges of sandy fields, recently burned-over areas, and other clearings. Another import from Eurasia, it is common from the Maritime Provinces of Canada to Alaska, and southward to New England, the Great Lakes, and the midwestern states.

Young leaves and stems of strawberry blite can be cooked and used just like lamb's-quarters. We like both species in our own cream of wild spinach soup. This involves first washing the leaves and boiling them until tender in water with a little vinegar added. For the soup we need a cup of cooked, chopped greens, so this requires quite a mass of plant material to start with, as it shrinks tremendously with cooking.

Once the cup of cooked, chopped leaves is secured, add 1½ cups chicken broth (or pheasant broth — that's the best), 1½ cups whole milk, and a small, finely chopped onion, all in a blender. Blend, then transfer to the top of a double boiler and heat slowly for 20 to 30 minutes. Next, season with a chicken bouillon cube (if desired), salt and pepper to taste, a dash of nutmeg, and a pinch of allspice. Serve piping hot with fresh crusty rolls and butter. This soup is very high in vitamins A and C.

The fleshy, strawberry-colored masses of fruit are also nutritious, and can be eaten both raw and cooked. However, they don't have very much flavor.

CHICORY
Cichorium Intybus

Chicory, or succory as it is sometimes called, is another wild plant that is hard to beat in a salad. Its popularity in Europe dates back as far as the Roman Empire, and even today the young leaves can be found in many marketplaces in both Europe and America. One can either harvest chicory leaves in the wilds in the spring of the year, or harvest the perennial roots and grow one's own supply in the basement. With all rays of light excluded from the growing area, the roots will send forth crisp, delicate white foliage.

It is a naturalized immigrant from Europe, and exists from coast to coast in most regions. It is very common in open fields, on roadsides, and in waste places. From July to October, where the plants grow in profusion, a sea of bright blue delights the eye when the most attractive

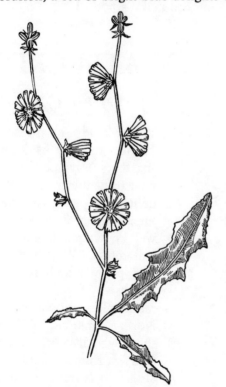

composite flowers are open. Not all chicory flowers are blue though. Two specific forms of the species sport white and rose-colored flowers. The flower heads are borne on leafed stalks, above the rosettes of leaves that arise from thick, strong taproots. The leaves are long, narrow, and have coarse teeth on their margins. Although they somewhat resemble dandelion leaves and are often collected indiscriminately along with the latter, c h i c o r y leaves are coarser a n d rougher in texture.

Young chicory leaves, in our opinion, make the best eating. We dig the young leaves of the rosette early in the spring, before they have developed too far and become bitter in taste from their milky sap. I have two favorite salad recipes. I tear up the greens and serve them fresh with olive oil and wine vinegar, or combine them with young spring onions, croutons, and a mild Italian dressing. Both are thoroughly enjoyable.

Once the leaves become older, they can be boiled in several changes of water and used as a cooked vegetable. However, the older the leaves get, the more bitter they become. Once the plant begins to flower, it is pretty hard to take.

Chicory has long been used as an adulterant or a substitute for coffee. It is available on most grocery shelves, but not many people know that it can be made from the roots of the weed with the pretty, blue, windmill-shaped flowers found in so many nearby fields. To make your own chicory coffee, dig a few roots, scrub them thoroughly, and roast to dryness in a moderately low oven. When the dark brown root is dried throughout, grind and use as a percolator or drip brew. Chicory coffee can be stored in any airtight container. Be careful, though; it's strong stuff, so go easy the first time you make a potfull.

WATERCRESS

Nasturtium officinale

Those cold, clear-running brooks and creeks harboring the subtly spotted speckled trout that make such a delicious pan-fried shore luncheon, also produce a well-known plant delicacy—watercress. Watercress is a very popular and somewhat expensive favorite in the marketplaces of Europe and North America. But for those who live near enough to a fast running, cold water stream, it is free for the taking as long as the water is open. There are, however, two precautions that should be observed when on a watercress collecting trip.

First, you must be certain that you are really harvesting watercress and not one of the less agreeable species that lives nearby. The poisonous water hemlocks grow in the same general vicinity, but their carrot-like form (described in the chapter on poisonous plants) is distinctively different from that of watercress, and virtually impossible to confuse with it if just a smidgeon of care is taken when collecting.

Second, watercress growing near heavily populated areas may be surviving in water that is not quite fit for consumption. So it is a wise idea to gather your watercress, bring it home, and let it soak in your kitchen sink for a while. Or you can purify your own water in the woods by dropping a halogen tablet (available at most drug stores) into a quart of water, waiting a half hour, and then washing your greens thoroughly.

Watercress grows most happily in the cool waters of brooks, spring-heads and rills, from Newfoundland to Ontario, south to Florida and Texas and westward through Iowa to California.

The plant is rather short, with somewhat of a spreading nature. The stems are creeping or freely floating in water, and sport from three to eleven roundish, deep green, glossy leaves. The last leaf at the tip of the stem is usually the largest. Small, delicate, white flowers give rise to long, needle-like seed pods known as siliques. Both the leaves and the young siliques are good to eat.

Watercress belongs to the mustard family, and exhibits the strong, sharp taste of the other mustard species. In my opinion this taste is best taken advantage of in salads. My family and I have a favorite meal that we indulge in at least once each summer that consists of thick, juicy mooseburgers done on the barbecue, fresh sweet corn from the garden, and watercress-onion salad. For the salad, thoroughly wash the watercress and leave it in a closed plastic bag with a couple of teaspoons of water in the crisper of the refrigerator overnight. This makes the cress as crisp as it was in the brook. Then finely shred fresh spring onions into a wooden salad bowl that has been rubbed with a cut clove of garlic, add the watercress, sprinkle with a few grains of salt and a few drops of oil and vinegar, and toss. Next add enough sour cream to moisten the cress, and toss again. This makes a marvelous salad with a lot of bite.

Watercress leaves can also be cooked and used as a table vegetable, but they shrink considerably while in the pot. They are good just boiled in salted water and smothered with butter, but some people prefer them mixed with more bland greens to cut down on the strong flavor. We like watercress leaves in soups and stews, precisely because of their sharp taste.

Another of our favorites is watercress butter. To 1 cup of soft butter, add ⅔ cup finely chopped watercress. Then 1½ tablespoons of lemon juice, ¼ teaspoon of Worcestershire sauce, and mix well. Chill slightly and serve on dark rye bread with poached eggs and hearty German sausage — one of my favorite brunches!

WINTER-CRESS

Barbarea spp.

Another member of the mustard family worthy of noting is the winter-cress, also known as the yellow rocket, herb of St. Barbara, and by a half dozen other names. The name St. Barbara, and indeed the genus name *Barbarea,* are derived from the seed of *Barbarea verna,* the early winter-cress, being sown in western Europe on or near St. Barbara's Day in mid-December.

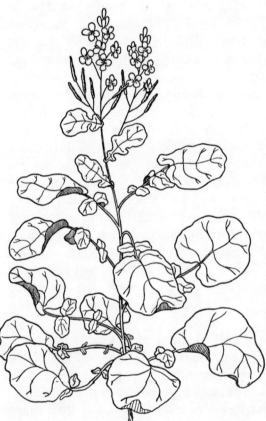

The name yellow rocket notes the bright yellow four-petalled flowers character-istic of all members of the winter-cresses. These flowers appear above and among the uppermost leaves. The tiny, cross-like flowers produce siliques that are somewhat four-sided in shape and fair-ly large in size. Below the flowers, in a rosette, appear stalks of smooth, green, rounded or oblong leaflets. The leaflet at the end of each stalk is the largest. The plants are biennial and per-ennial in nature, with the basal rosette surviving the winter.

The common winter-cress, *Barbarea vulgaris,* is a naturalized citizen of meadows, damp woods, and brook and stream banks. Its cousin, the early winter-cress, often grows side by side with it in weedy fields. Between the two species, they cover an area ranging through most of southeastern Canada, New England, the Atlantic seaboard to Florida, the Great Lakes and midwestern states, and on to California.

As the name indicates, winter-cress leaves are among the first to appear in very early spring. They even have a habit of poking up during mild spells in winter. Young leaves are quite delicate in flavor, but become stronger as the year's growth matures. Once the plants are in flower, the leaves have quite a bite, and are best handled as a cooked vegetable in two changes of water. The buds, harvested before the flowers open, also make a good potherb. I boil them and top with a rich, thick, wine and cheese sauce, made by adding ⅓ of a pound of shredded mild cheddar cheese and ⅓ of a cup of Spanish burgundy to a basic white sauce.

The best way to enjoy young, crisp winter-cress leaves is in a spring salad. As with watercress, make liberal use of shredded spring onions, both green and white portions, and sour cream. A bit of crumbled Roquefort cheese in this salad also adds a nice touch.

MUSTARD

Brassica spp.

Of the five species of mustard that abound throughout the temperate regions of North America, the black mustard, *Brassica nigra,* is the best known. Naturalized from Eurasia, it has indeed become a useful citizen. It is our principal source of table mustard.

Black mustard is an annual. It ranges from two to seven feet in height, and its branches spread voluminously. The leaves on the lower parts of the plant bear tooth-like lobes, but these disappear as you progress up the stem. The leaves that appear on the flower stocks have almost smooth margins. The bright green foliage of young plants sports hairs and fuzz over the entire leaf surface.

A relative of turnip, cauliflower, Brussels sprouts, and other cultivated cole crops, black mustard grows in every nook and cranny of waste

space and in cultivated fields from coast to coast. Almost everyone recognizes t h e acres of sunshine that it provides when its multitudes of bright, tiny, yellow flowers are in full bloom. As with all other crucifers or mustard family members, black mustard flowers sport four petals in two pairs, with six stamens — four long and two short. The flowers produce short seed pods filled with pungent, dark seeds.

It is from the seeds that table mustard is prepared. You can collect mustard seeds in late summer, bring them home and let them dry out thoroughly in a warm place for forty-eight hours, and then grind them finely (a fine pepper grinder or a blender does a good job) to make a dry mustard that I think is far better than the stuff you buy in a tin. You can make your own table mustard by dissolving the dried mustard material to make a paste with two-percent milk.

To make mustard sauce, mix 2½ tablespoons of dried, ground mustard seeds, 1 teaspoon of flour, a dash of salt, and ¼ cup of cereal cream. Then add this mixture to another ¾ cup of cereal cream and ¼ cup of sugar that have been heated. After stirring the two together and heating again, add 1 beaten egg yolk and stir once more. Then put it back on the stove until the sauce thickens. After thickening, add ½ cup of heated white vinegar, and you have got a superb and zesty sauce for ham, fish, beef, or a number of cooked vegetables. We particularly like it on some of the market fish that we get in winter. Somehow these creatures just don't have the flavor of our brookies and rainbows, and the mustard sauce perks them up nicely. Whole black mustard seeds are also a favorite ingredient in Indian curried dishes.

We cannot, of course, forget mustard as a fresh green, because that is probably when it is at its best. The leaves of very young plants are the ones to pick. Their flavor has a distinct peppery tinge to it, and makes a fine salad with other spring greens and a mild Italian dressing.

The flower buds and golden-colored flowers can also be eaten. As they are much like broccoli, we enjoy a mild cheese sauce over them. We also like them boiled and drained, smothered in butter, covered lightly with a mixture of seventy-five percent shredded cheddar cheese and twenty-five percent Romano, and tucked under the broiling element for a few minutes.

As if all this were not enough, young mustard leaves can be boiled and used as a potherb. They also make a delicious cream of mustard soup. Even when the leaves become older, we use them in soups and stews, where their pungent flavor is really appreciated.

SHEPHERD'S-PURSE
Capsella Bursa-pastoris

Another member of the mustard family, shepherd's-purse, or pick-pocket as it is sometimes dubbed, combines its lightly peppery taste with a high nutritive value, to make a choice spring green for the the wild harvester. It is good both cooked and raw, again becoming stronger in taste as the plant's parts get older. Late winter and early spring are the times to collect young leaves. After that they toughen considerably. In summer and autumn the seeds ripen, and various Indian tribes are said to have collected them to dry and grind into a nourishing meal.

Shepherd's-purse is a very familiar garden weed. Its leaves form a rosette arising from the roots. The leaves of the rosette are a mixed lot — all elongated, but some bearing coarsely toothed margins and some being entirely smooth-edged. The plant produces long clusters of white flowers, with the two opposite pairs of petals characteristic of the mustards. These flowers mature to form flat seed pods that are triangular in shape, and suspend themselves from the plant with the base of the triangle upwards. This latter is the characteristic that is easiest to recognize.

Aside from gardens and lawns, which are both favored living quarters of the pickpocket, the plant grows along pathways, roadsides, in cultivated fields, and in all waste places where civilization exists nearby. It grows over the whole of North America, and indeed over a good portion of the rest of the world.

Shepherd's-purse is a bit tricky in the salad bowl. It tends to shed whatever type of salad dressing you try to put on it — even the gooey ones. I have had the best luck with drying the leaves thoroughly after washing; then tossing the torn parts vigorously with an oil and vinegar dressing until the dressing sticks. Serve immediately before the dressing gets away!

I have tried pickpocket leaves cooked in various ways, and I prefer them in boiled dinners, along with potatoes, carrots, turnips, onions, and a big ham bone. They lend a nice peppery flavor to stews as well.

Lastly, dried and ground shepherd's-purse seeds make a tasteful mustard bread, when mixed 1 to 2 with corn meal. The taste is quite strong, and it's a lot of work, because after the grinding operation is over, the vacuum cleaner has its job of retrieving the many small seeds from every nook and cranny of the kitchen.

Shepherd's-purse tea is said to be a stimulant for sluggish kidneys, and the pioneers used it as a dressing for bruised limbs. They are said to have soaked the leaves in water and used them as a medicinal wash.

ALPINE CRESS
Arabis albina

Alpine cress, as the name indicates, is one of the far northern members of the mustard family. The plant is seen in rocky places — cliffs,

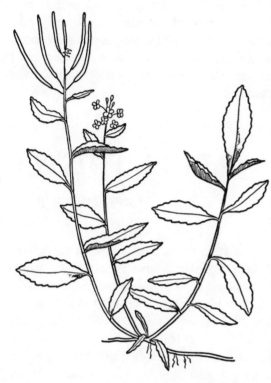

ledges, roadsides, s t r e a m banks, especially gravelly shores, and cool rocks. It is also found cultivated in rockeries. Its range extends over most of the eastern Arctic and as far south as western Newfoundland and Quebec's Gaspé Peninsula.

Alpine cress is a matted perennial. Its loose rosettes are comprised of fleshy leaves that are coarsely toothed and hairy on both surfaces. The leaves average about an inch in length. The flowering stems are loose, from a few inches to a foot or so in length, and bear a few leaves that arise almost directly from the stalk. Long, loose clusters of white flowers adorn the plant in spring and summer. The four-petalled blooms are less than a half-inch long, but extremely pretty.

Alpine cress is best eaten as a fresh salad green, but can also be used as a potherb. The leaves of the rosettes make the best eating. It has a fairly pungent taste, similar to radish (which is yet another mustard relative). Some people prefer to dilute the strong flavor by combining it with other, milder greens before pouring on the oil and vinegar. My family like it straight.

Once, during a photographic expedition far into northern Quebec, we were fed an alpine cress potato soup. The memory of that soup lingered until long after our return, and the next time we were in the Gaspé, we tried it again. It is truly a delicious soup, and if you don't happen to live near the eastern Arctic, try it with one of the other *Arabis* species, the rock cresses, which abound over a good portion of the cold part of this continent.

Our potato soup involves boiling 3 cups of potato chunks, 2 cups of chopped celery, 1 cup sliced onions, and salt and pepper, along with a ham bone until the vegetables are tender. In the meantime, mix ¼ cup of flour with 4 cups of milk. Remove the ham bone temporarily and add the milk mixture to the potatoes. Then add a couple of tablespoons of butter and about 4 cups of shredded alpine cress. Stir and heat until the soup has thickened, return the ham bone, and simmer until all the flavors mix. This soup really sticks to the ribs.

BULL THISTLE
Cirsium vulgare

I can remember some years ago a student earning a few extra dollars by collecting materials for a very popular course in arranging dried flowers. Lots of wild material was used, and plants were collected in vast quantities. This chap, armed with a pair of pruning shears and a list of the species required, set off one weekend on a collecting spree. At the top of the list was "thistle" — or at least that's what he thought. So he trudged around pastures and along roadsides, quickly becoming nicked and cut and scratched by the relentless bull thistles, and finally secured enough to fill his quota. Monday morning he proudly presented his masses of thistles, so painfully procured, to the course's instructor, who immediately informed him that he must have misread her handwriting — it was teasel she wanted, not thistle! Upon later hearing the story from the thistle collector, who still bore the scars of the expedition, I informed him that the devilish, prickly beasts he had searched out were an edible species, and quite a tasty one. It was at that point that he determined I had gone completely off the deep end!

Thistles don't really look terribly appetizing, and indeed you may not think that fighting the prickles to collect and prepare the plants for the pot is worth the end result. There is no doubt that a field trip to collect bull thistles requires arming oneself with decent boots, scratch-proof denim or cotton twill field clothes, and a pair of gardening gloves.

The bull or common thistle is a biennial. In its first year of growth, it produces a rosette of coarsely toothed, dandelion-like leaves with woolly hairs on the undersides and long, sharp prickles on the upper surfaces. The flowering stalk, formed in the second year, is one to six

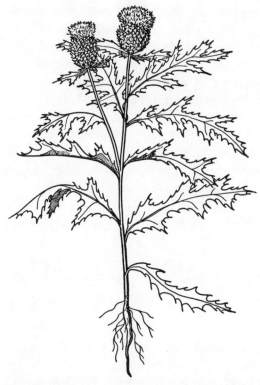

feet in height. It bears leaves of the same nature as the rosette leaves, but considerably smaller. The stem also has prickly projections.

Like all members of the composite family, the common thistle has flower heads composed of many small flowers. The flower heads, one to two inches high and purple in color, can occur either singly or in clusters.

The bull thistle is an aggressive weed. It travels and establishes itself wherever it can find a suitable clearing, pasture or roadside. In many areas it is a troublesome weed, competing with more desirable species, and making pasture land rough and unusable. A native of Europe and Asia, it has pushed its way from coast to coast in North America, throughout the temperate zone.

The thistles are excellent emergency foods because they are so easily recognized. The stems, stripped of their thorny projections, can be eaten raw or cooked. The leaves can also be eaten — minus the prickles.

The Navajo Indians of the western states are reported to have chewed fresh roots of the carmine thistle, *Cirsium rothrockii,* as a treatment for smallpox. They also prepared a lotion of the plant to treat the same illness. The Kiowa Indians reputedly used roots of the yellow-spined thistle, *Cirsium ochrocentrum,* as a foodstuff, and used a lotion made of the boiled flowers as a burn remedy. One Indian tribe of the Pacific northwest is reported to have drunk thistle tea as a contraceptive measure. And the Zuñis drank a cold thistle tea as a treatment for syphilis.

Young thistle leaves make delicious salad greens with chunks of fresh tomatoes, bits of artichokes previously preserved in oil, and fresh carrot curls. For this salad, we use a dressing made as follows. Place in your blender jar ½ cup of olive oil, ½ cup of vegetable oil, ¼ cup of wine vinegar, freshly ground pepper, and the juice of one crushed clove of garlic. Blend at medium speed for about 10 seconds. Then add the cooked yolks of 4 eggs. Our method of doing this is to separate 4 fresh eggs, use the whites for a meringue, and then cook the 4 yolks with a little butter in the egg poacher. Finally, add ¼ of a sweet pepper and one small onion, finely chopped, and blend until smooth. This dressing will keep in a tightly closed container in the refrigerator for a couple of weeks.

As a cooked vegetable, the stem of the thistle is the easiest to prepare. It should be peeled before cooking, and cut into convenient lengths Boiled just until tender, it is good served with a sprinkling of salt and pepper, and butter.

SOW THISTLE
Sonchus spp.

The sow thistle is a thistle of a different ilk. As a member of the genus *Sonchus*, it is more closely allied to the dandelions than to the other thistles. Indeed, it strongly resembles a prickly dandelion, and can be treated in much the same way for table use.

A with the bull thistle, the sow thistle should be collected with the aid of a pair of gardening gloves. Young leaves are the easiest to handle, and as the leaves become older you may want to cut them so that they drop directly into a bag or basket without any excessive handling.

The flowers of the sow thistle are easy to recognize. They grow in clusters of a few to many, at the summit of the flowering stalk. They are composite in nature, and each flower head averages less than an inch in height. Yellow in color, the flower heads resemble small dandelion flowers. The seeds of the sow thistle are the favorites of many passerine birds, including some of the finches. The leaves, which essentially look like bristly dandelion leaves, arise almost directly from the stem.

Naturalized from the Old World, the sow thistle has essentially run rampant in North America. It occurs from coast to coast, throughout the temperate regions. It is basically a lover of the civilized world, and abounds close to populated areas in cultivated fields, on roadsides, in

barnyards, and in many wastelands. Over a good portion of the continent it is considered a troublesome weed.

When preparing sow thistle leaves for the table, one must remember that as a close relative of the dandelion, the plant contains a milky sap that is somewhat bitter in flavor. If you enjoy a bitter taste, you can use the young leaves in a salad and the older ones as a cooked vegetable, simply boiled in salted water and smothered with butter. If you are not an advocate of the bitter flavor, you may want to boil the leaves in two or more changes of water. The more changes, of course, the milder the end result will be.

Young sow thistle leaves make a nice Caesar salad because of their mildly bitter taste. I also toss them with chunks of tomatoes, spring onions, thinly sliced European seedless cucumber, finely chopped celery and white radish, chopped olives, and seasoned croutons. An oil and vinegar dressing, with a dash of dry mustard and Worcestershire sauce, completes the salad.

LIVERBERRY
Streptopus spp.

The liverberry has several interesting names. This first one is indicative of the cathartic properties of the plant's scarlet-colored berries. To the country folk of New England, the plant is known as scootberry, for

somewhat imaginable reasons. The liverberry, though, is most commonly known by the name of twisted-stalk, a title most aptly describing the growth character of the plant. Even the genus name *Streptopus* comes from the G r e e k *streptos* meaning twisted and *pous* meaning foot. However, to the wild harvester, this plant can only be known as wild cucumber, because it is one of the few wild edibles with a taste truly like our domestic cucumber. Of course, it is not the berries of the wild cucumber that make up our salad (that might be somewhat disastrous), but rather the young stalks of the plant.

Twisted-stalk is a perennial member of the lily family. It grows to about four feet in height, above a thick, short rhizome. The stems can be either single or branched, and are clasped on alternate sides by bright green leaves. These leaves are longer than they are wide, and sport numerous parallel veins running prominently down the length of the leaves.

The single, tiny, greenish-white to pinkish flowers hang like miniature bells from small stalks arising underneath the leaf axils. These flowers give rise to pulpy, pendulous berries, ranging in color from pink to deep scarlet. The berries have a mild cucumber-like taste.

The liverberry is an advocate of moist woods and thickets, including peat areas and small openings. It thrives across the northern portion of this continent, from Labrador and Newfoundland to the West Coast, and is very common in New England, the Canadian Maritime Provinces, and in the Great Lakes area.

Young shoots are best enjoyed in a wild cucumber salad. Our favorite method is to tear up the young liverberry shoots and place them, along with a finely shredded Spanish onion, in a bowl with an airtight lid. Then salt and pepper to taste and add about ⅛ cup of mild white vinegar. Put the lid on the container, shake vigorously to coat everything with the vinegar, and leave in the refrigerator for a couple of hours. About 15 minutes before serving, add enough sour cream to the wild cucumber and onion mixture to coat everything lightly, and then shake again. This is a marvelous cottage lunch when combined with fresh yellow perch fillets or the previous evening's catch of brown bullhead.

We also make a creamed soup of wild cucumber shoots, and although it is a little time-consuming, it is well worth it when the steaming hot dish appears on the table. For this you need a quantity of the young shoots finely chopped. First, chop a medium-sized onion and simmer it in butter until tender and translucent. Then add the greens, cover, and simmer until the greens are cooked, but not overdone. Add about 3 tablespoons of flour to the hot vegetable mixture and stir to thicken. Then slowly add 2 cups of chicken stock. In a separate pot, scald 1 cup of milk, and add it to the rest of the ingredients. Next, pour the whole soup mixture into your blender jar and blend until smooth. If passing the mixture through a sieve is more convenient, then by all means do it that way, although I find the blender easier.

Put the soup into a heavy pot, add a few grains of mace, salt and pepper to taste, and bring it to the boiling point. Just before serving, add ½ cup of cream and serve piping hot. This soup has a very distinctive flavor and goes down very well before a meal of roasted poultry.

INDIAN CUCUMBER-ROOT
Medeola virginiana

While we are on the subject of wild edibles that taste like cucumbers, there is one whose beauty surpasses even its delicious flavor. The Indian cucumber-root, or more commonly just Indian cucumber, is one of the most attractive plants of the eastern woods.

Another member of the lily family, the Indian cucumber inhabits rich soil woods areas from south-central Canada to Minnesota, and from the Canadian Maritime Provinces through New England, and

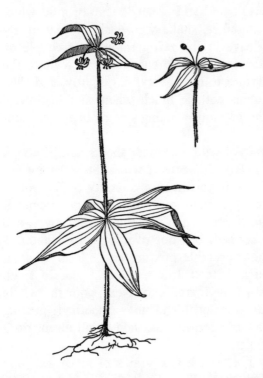

south to Florida and Louisiana.

This perennial herb has a single, slender, erect stem, arising from a horizontal tuber, and the stem is covered with cobweb-like hairs. From half to two-thirds of the way up the stem, five to nine leaves form a whorl around the stem, drooping ever-so-lightly like a graceful, green skirt. At the top of the stem sits another whorl of three (or rarely more) leaves, giving the plant a bright, sporty collar. In fall, the leaves take on a distinctly purple hue.

The flowers of the Indian cucumber-root are spider-like in shape, and the color of strawberries. They arise from the axils of the upper whorl of leaves. The berries produced by this plant are dark purple to almost black in color. Their three-celled interiors harbor very few seeds.

The most important part of the plant as far as its edibility is concerned is the tuberous rootstock. This crisp, white underground portion of the plant is about the size of a man's small finger, and possesses a flavor that is very close to that of the domestic cucumber. It can be harvested in spring, summer, or fall, depending on locality, but the conservation-minded wild harvester has to remember not to decimate the supply of the species in any one area. The delicious flavor of the tuber tends to encourage extensive picking if a patch of reasonable abundance is found, but remember to keep your harvest to the overflow of the population. With most plant species, a little judicious pruning won't hurt, but try not to overdo it.

Indian cucumber rootstocks can be cooked, but are best enjoyed as a salad ingredient, thinly sliced and tossed in with small pieces of fresh leafy lettuce. A simple oil and vinegar dressing best brings out the cucumber flavor.

What domestic species makes better pickles than the cucumber? Since my family couldn't think of any, we decided it was time to explore the possibilities offered by wild cucumber pickles. Then, because we didn't have too many of the crisp, white tubers, we decided to alter our plans slightly to produce wild cucumber and onion pickles. For this, wash and cut into large chunks enough cucumber-root tubers for a little over 1 cup of material. Then cut a large, sweet Spanish onion into chunks of the same size. Sprinkle the two combined with a little salt, and let them stand to partially dehydrate for a half hour. In the meantime, make a pickling brine with ¾ cup mild white vinegar, ¾ cup sugar, 1 teaspoon celery seed, ½ teaspoon dill seed, ½ teaspoon mustard seed and ½ teaspoon turmeric. Drain the wild cucumber-onion mixture and toss it all in with the brine, slowly heating until the brine is hot. Pour into two small canning jars and refrigerate. Our original intentions were to leave our pickles for three or four months before sampling. But of course that never worked out. The first taste testing started about two days after the original canning operation, and the pickles had been completely demolished by the time the third month came around.

Although the generic name *Medeola* is said to be a namesake of the sorceress Medea, because of imaginary medicinal powers, the Indian cucumber doesn't seem to have any history of use as a medicinal plant.

MINER'S LETTUCE
Montia spp.

Miner's lettuce is another of the plants steeped in early American history. It was the salvation of many a gold miner whose complete absorption with the prospect of finding new riches left him little time to concern himself with such mundane matters as nutrition. Consequently scurvy, the vitamin deficiency disease of explorers, also ran rampant among miners. Too busy to bother with the planting and maintenance of a cultivated patch of greens, miners took to eating the wild lettuce that was introduced to them by the redskins of the west. The

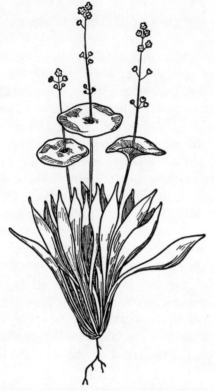

name miner's lettuce is a vestige of the days of California's famous Forty-Niners who harvested the wild edible to stave off scurvy in their prospecting camps.

Miner's lettuce is a western plant. Also known as Indian lettuce, Spanish lettuce, and in Europe as winter purslane, this member of the purslane family is found throughout the greater part of western North America from British Columbia to Mexico, and eastward as far as North Dakota. It is an inhabitant of moist soil areas, and is common on shaded slopes and river banks, in cultivated fields and vineyards, and in pastures.

Miner's lettuce is quite easily recognized, even by the novice. It has basically two types of leaves — long, slender ones on leaf stalks arising from the base of the plant, and rounded upper ones that are most characteristic of the plant. The upper ones occur in opposite pairs, joined to form a skirt around the stem, with the stem continuing through the joined leaves and terminating in a flowering stalk. In addition to these, other leaves on the plant resemble fleshy triangles or assume the shape of a kidney. All of the leaves and stems are edible, but the long, slim basal leaves tend to be a bit tough.

The flowering stalks of miner's lettuce sport tiny, loosely bunched flowers that vary from white to pinkish in color. In autumn, shiny black seeds abound on the plant.

Being from the East, I spent the first two decades of my life completely unacquainted with the delicacies that can be produced from Indian lettuce. That thankfully changed one fine early summer day when I was introduced to a miner's lettuce salad made by tossing the green leaves with small chunks of tomatoes, crumbled Roquefort cheese, and an oil and vinegar dressing. It was a delightful experience, followed the next night by a similar salad dressed up with croutons and thinly sliced sweet onions.

Once early summer has started to age and miner's lettuce leaves become not-so-young, they can be used in almost any spinach recipe. One Sunday afternoon we decided to try an Indian lettuce soufflé — with marvelous results. We started out with quite a batch of lettuce leaves and simmered them in just a shade of water until they were tender. Then we drained and chopped them finely. To 3 cups of the cooked miner's lettuce, we added 2 heaping tablespoons of butter, 3 beaten egg yolks, 1½ cups of whole milk, ¼ teaspoon sugar, salt and pepper to taste, a few grains of ground mace, and 1 teaspoon of freshly squeezed lemon juice. We mixed all these ingredients thoroughly, and then folded in 3 egg whites that had been beaten until they were stiff. This mixture was poured into a buttered soufflé dish (although any non-stick casserole dish will do) and baked in a moderately slow oven (300° F) for about a half hour. The finished product was firm, served immediately and cleaned from everyone's plate just about as fast.

The Indians of the Pacific Coast are reported to have seasoned their Indian lettuce before eating by piling it high over the nests of red ants and allowing the ants to crawl and climb through it. Then they shook the ants out and ate the product. It sounds quite intriguing, but somehow we would rather just stick to wild lettuce soufflé!

SCURVY GRASS
Cochlearia spp.

I can well remember the vivid and rather morbid history book descriptions of a debilitating disease that struck pioneer settlers, explorers, prospectors, trappers, sailors, and other adventurers alike. The pain and suffering inflicted by this lethal disease were devastating. The cure was a simple one, but it took some years to discover. Thus thousands and thousands, in the days before refrigeration facilities and chemical drug preparations, died at the hands of the killer known as scurvy.

Scurvy is a vitamin deficiency disease. Its cure, and indeed its prevention, are easily accomplished by providing the body with a sufficient supply of vitamin C. This is an easy matter today. But the early

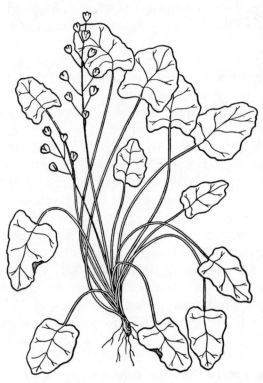

explorers and prospectors didn't have the orange-flavored chewable tablets that we make our kids swallow every day. They had to rely on green plants to supply their vitamin C. One of the wild edibles that they utilized in great quantity, because of its high vitamin content, was *Cochlearia* — appropriately named scurvy grass.

There are numerous accounts of seamen in the early part of the nineteenth century gathering g r e a t quantities of scurvy grass and loading it on board their ships in bales. The greens helped to prevent sailors from contracting the dreaded disease. How long it was effective is hard to say. As with today's fruits and vegetables, the vitamin content of scurvy grass rapidly decreases as the cut greens become old.

Scurvy grass is another member of the mustard family. It tastes something like watercress, and has the strong odor of horseradish about it. Three species of scurvy grass grow in North America, mostly in rich, damp, brackish or calcareous soils near the coast. The range of the plants encompasses the northern portions of the continent, from Newfoundland to Alaska, including the many islands in the Gulf of St. Lawrence.

As with many of the mustard family members, scurvy grass is a biennial. Its leaves are somewhat spoon-shaped, and indeed the name *Cochlearia* is derived from *cochlear* meaning spoon. The leaves are mostly supported on slender stalk, and form a rosette a few inches in diameter. Both smooth and toothed leaf margins occur. The fleshy leaves are almost veinless.

The small white flowers of scurvy grass have two pairs of petals, arranged in opposite pairs, like the cross that is characteristic of all of the mustards. The flowers grow on branched stems, a few inches high, arising from the old rosettes. From the flowers, flattened, oval-shaped seed pods develop.

Scurvy grass can be harvested when the leaves are young. They make delicious eating when fresh. The young, lower leaves can also be taken, even when the plant becomes older. As long as the leaves and stems are tender, the eating is good. Raw or cooked, scurvy grass offers a variety of dishes that are healthy and nutritious.

I must say that we prefer the raw leaves to the cooked product. The spicy flavor of the leaves is readily enjoyed in salads, either solo or mixed with other, less pungent leafy greens. A light oil and vinegar dressing lets the horseradish flavor through nicely. We like a salad made of scurvy grass leaves and slices of hard-boiled eggs, with small pieces of sharp cheddar cheese thrown in.

Scurvy grass is pretty hard to beat in a sandwich, either by itself or combined with other fillings. One member of our family is an avid supporter of egg salad sandwiches made from chopped, hard-boiled eggs, finely diced mild onions, and chopped scurvy grass. I prefer ham and cheese sandwiches on heavy rye bread, stuffed with a liberal amount of the vitamin C grass. The sharp flavor perks up the sandwich considerably.

An idea that came to us via a friend several years ago was *Cochlearia* salad dressing. There are two versions — the "from scratch" and the "quickie." For the long version, beat one egg, ½ teaspoon of dry mustard, ½ teaspoon of salt (or more if you are a salt addict), and a dash of ground hot pepper, with a portable electric mixer or a wire whisk. When the mixture is smooth, begin adding ¼ cup of vegetable oil, a few drops at a time, beating all the while. When the mixture is very thick, start adding 1 tablespoon of freshly squeezed lemon juice, still beating. Then add another ½ cup of oil, 1 tablespoon of lemon juice, and a final ¼ cup of oil. You must beat this mixture religiously during all of the additions to keep it smooth. Now add ½ cup of finely chopped scurvy grass leaves and refrigerate. We tried this recipe several times, and the dressing was so popular that we developed our "quickie" method, which is much easier and, we think, almost as good. To 1 cup of commercially made mayonnaise (one with lemon juice added), we add ½ cup of chopped greens. Presto — instant scurvy grass salad dressing.

Scurvy grass can also be enjoyed cooked — just simmer for a few minutes and serve with salt, pepper, and butter. It also makes good soups and stews.

DOCK

Rumex spp.

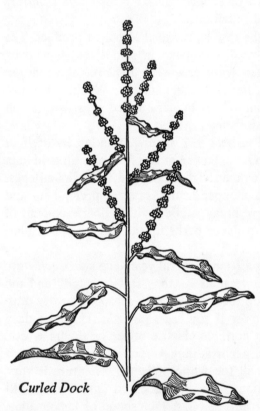

Curled Dock

The genus *Rumex,* of the buckwheat family, comprises two main groups of plants — the docks and the sorrels. We will first consider the docks, all of which are edible. The various species of dock vary considerably in size of the plant and in details of the leaves and flowers. However, once you have learned to recognize them, you will quickly become a dock expert.

The yellow dock, *Rumex crispus,* is steeped in ancient medicinal history. It is reported to have been very popular with early American redskin medics in drawing out the "poison" in boils and other festering cuts and sores. The root was also used as a laxative, an astringent, and for a half dozen other medical tasks. How effective the plant was in its curing powers is not known. Certainly it helped in a lot of cases where the cause of the ailment was a lack of vitamins A or C. For example, many people suffering from loosening teeth caused by vitamin C deficiency, which softens the gums, benefited from the consumption of dock greens. Also people suffering visual difficulties directly caused by a deficiency of vitamin A were cured by an intake of dock leaves.

Approximately fifteen species of dock exist on this continent. All of these species are medium in height and quite stout. The largest leaves occur at the base of the plant, where they may grow from six inches to almost two feet in length. The leaves further up the plant's stem are smaller. The leaves are tapered from their middle, sometimes rounded or heart-shaped, and have smooth undersurfaces. At the end of the petiole, where the leaves attach themselves to the stem, appears a thin, membranous sheath which wraps itself around the stem.

The flowers of the dock are rather disappointing. They are light green to purplish in color, and are somewhat insignificant except that they grow in long, wand-like clusters. The fruits are much more interesting. They bear numerous small seeds, each sporting thin, ribbed wings. As one watches the flowering stalks of the dock from early summer to late fall, one can see a gradual and most impressive change from all shades of green, through pinks and tans, to rusts and browns.

The docks abound on roadsides, in ditches, in fields and pastures, and in many waste areas throughout the temperate regions of this continent. They occur sometimes singly and at times in great clusters. Their range extends in the west from Alaska and British Columbia through all of the Pacific states, and in the east from South Carolina north to southern Labrador. Several species also occur in swampy areas.

Dock can be harvested at any time of the spring or early summer before the leaves become tough. After that, they develop a strong bitterness, along with a sour taste. However, once a frost has hit in the fall, some dock species rejuvenate enough to produce some new foliage, hence the plant can be harvested at that time as well.

Dock is reported to be higher in vitamin C than citrus juice and richer in vitamin A than carrots. Indeed, young dock leaves have something of a citrus taste about them, so in a salad they should be topped with a salad dressing that is light in vinegar or lemon juice. Mixed half-and-half with dandelion greens or leafy lettuce, they make a good spring salad, especially with slivers of smoked ham and tiny bits of sweet pickle. Cover with a dressing made of 1 cup corn oil, 1/6 cup mild vinegar, 1 teaspoon sugar, 1 teaspoon sweet paprika, 1½ teaspoons dry mustard, and salt and pepper to taste. The mustardy taste of the dressing goes well with the ham and the slightly bitter greens.

Young dock leaves can be simmered slowly in water for use as a cooked vegetable. Older ones will need two or three changes of water to lessen the bitter taste. We like our cooked greens smothered with butter and sprinkled with grated Parmesan cheese. Often after boiling dock greens, we chop them and toss them in a frypan with onions and a touch of garlic that has been sauteed until tender. Sometimes we add a dash of nutmeg or mace to the greens, instead of the Parmesan cheese.

SORREL
Rumex spp.

Sheep Sorrel

Those members of the genus *Rumex* that are not docks are sorrels. In fact, the two groups of plants resemble each other quite closely, except that the sorrels are usually smaller than the docks and bear arrow - shaped leaves. The sour taste of the docks persists in the sorrels as well.

The common or sheep sorrel, *Rumex Acetosella,* is probably the best known of the few sorrel species. It is a fairly short plant, stretching from eight to twelve inches above its slender, running rootstocks. The leaves of the plant are light green in color and somewhat resemble the leaves of *Sagittaria* because they are shaped like an arrowhead or spearhead. The flowers occur in spikes at the top of the plant, and often assume a reddish tinge. This reddish color is also seen at times in the roots and on the leaves. Individual plants bear either male or female flowers, not both.

The sheep sorrel is a resident of fields and gardens that have long passed their prime. It thrives in soils that are dry and sterile or sour.

Naturalized from Europe, it has settled across a great portion of North America.

The garden sorrel, *Rumex Acetosa,* is a coarser, stouter plant, stemming from a tough taproot, and with ribbed, succulent stems. It inhabits all manner of fields, ditches, roadsides, and meadows from Greenland and Labrador to Alaska and the western coast of Canada, and south in the temperate zone. It, too, is an immigrant from Europe.

The sorrels are known to make excellent salads. Indeed the garden sorrel was at one time a very popular salad ingredient, but has fallen over the years into disuse. The young leaves and sprouts (long before the flowers develop) are the best, and are sometimes preferred blanched. This can be accomplished by growing sorrel in a dark place or by putting a paper bag over it while it is growing. The lack of light will cause a degradation of the plant's chlorophyll, and the resulting near-white shoots and leaves make an interesting tossed salad.

Because of the sour taste, sorrel salad should be topped with a dressing containing very little lemon juice or vinegar. In fact, many people prefer to dilute the sourness by mixing sorrel greens with other leafy greens, spring onions, and tomatoes. We use the same mustardy dressing on sorrel that we do on dock.

Sorrel makes very interesting soups (both hot and cold), purées (to be eaten on toast), and lends a nice bite to boiled dinners — the kind with a big ham bone in the bottom of the pot. Another interesting sorrel dish can be made by cooking enough greens in salted water, and draining them to give 3 cups of cooked, chopped greens. Melt about 1/5 pound of butter in a pan and add the greens to it. When the greens are heated through, sprinkle with 2 tablespoons of flour and add ⅓ cup cream and ⅓ cup chicken stock. Then season with salt, pepper, and a dash of sugar. Transfer to a serving plate and garnish with seasoned bread crumbs.

The potassium oxalate content of sorrel also makes for an interesting and thirst-quenching drink. For this, make a thin syrup by using about 6 parts water to 1 part sugar. Then add 5 or 6 cups of shredded sorrel greens and bring to a boil. Remove from the heat and let the mixture steep for about 3 hours. Strain the sorrel greens from the drink and cool the liquid in the refrigerator until ice-cold. Then mix half-and-half with sparkling mineral water or club soda. If you like the drink a little stronger, add more sorrel leaves at the beginning of the batch.

MOUNTAIN SORREL
Oxyria digyna

The mountain sorrel, despite its common name, is not a member of the genus *Rumex*. It is, however, very closely related, and has a similar sourness in taste to the docks and the other sorrels. This is evidenced in the name *Oxyria* which is derived from the Greek word *oxys*, meaning sour.

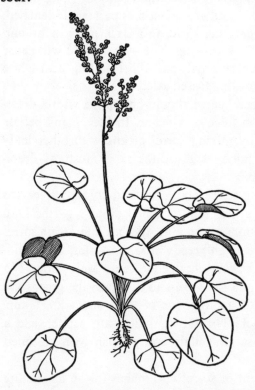

The mountain sorrel is a short, arctic and alpine species. It commonly reaches only six to eight inches in height, and rarely goes to two feet. A perennial member of the buckwheat family, the plant bears one and sometimes two leaves on each "stem," really a leaf petiole generally arising in tufts directly from the rootstock. The leaves are rounded or assume the shape of a kidney. Atop the succulent, stoutish stems are spires of tiny, insignificant greenish to crimson flowers. These flowers mature to form small, rounded capsules that are reddish in color and sport a thin, veiny wing.

A native of Eurasia, the mountain sorrel has settled in many damp, waste places in the arctic regions from Greenland to Alaska, in the rocky slopes of Newfoundland and Eastern Quebec, west through the alpine regions to the Rocky Mountains, and south as far as New Mexico, Arizona, and southern California.

The plants are best presented on the table in late spring or early summer, before the flower clusters appear. After that, their sour taste is very prominent and the leaves become somewhat tough. Mountain sorrel has a taste vaguely resembling that of rhubarb, and we have seen it used with success in some rhubarb dishes, although the texture and consistency of mountain sorrel leaves and rhubarb stems is completely different. (One should take note here that domestic rhubarb leaves, both raw and cooked, are poisonous.)

Mountain sorrel greens, also known in various localities as scurvy greens, sour grass, and alpine sorrel, make a delicious salad. Mix the torn greens with leafy lettuce curls (for color), celery curls, finely sliced spring onions (including the green sections), crumbled bacon bits, and small pieces of sharp cheddar cheese. A good salad dressing for this is made by stirring 2 tablespoons of flour and ½ finely minced onion in a heavy saucepan containing 4 tablespoons of bacon fat. Stir constantly over low heat for a couple of minutes. Then add 1 teaspoon prepared mild mustard, 1 teaspoon prepared hot mustard, ¼ teaspoon sugar, salt and pepper to taste, and blend well over the stove. Then slowly add 1 cup of water and stir until thick. When cool, pour over the tossed salad. The sour taste of the mountain sorrel comes nicely through the mustard and bacon dressing. Incidentally, we like the flavor of mountain sorrel and don't bother diluting it with leaf lettuce.

As a cooked vegetable, mountain sorrel holds its own. The young leaves can be boiled by themselves or combined with other greens. Topped with butter or sautéed onion and bacon bits, this dish is great with lake trout fillets.

We have a favorite mountain sorrel and watercress soup made by combining 3 tablespoons of melted butter, 2 tablespoons of flour, 2 cups of mountain sorrel greens (packed), 2 cups of watercress greens (packed), and 4 cups of game bird broth in the blender container. Then buzz around until smooth. If all this won't fit in your blender at one time, you will have to do the buzzing in two batches. Then heat slowly to the boiling point. Cool slightly and add 1 teaspoon of sweet paprika, a dash of pepper, and 1 cup of cereal cream. Heat again. Pour into a large tureen, garnish with seasoned croutons and watercress sprigs, and serve right away. This is a soup that always calls for seconds!

WILD LETTUCE

Lactuca spp.

Numerous species of wild lettuce are known to those who seek edible wild plants. However, two of the most familiar are the common wild lettuce, *Lactuca canadensis,* and the prickly wild lettuce, *Lactuca scariola.* Both are close relatives of the domestic garden lettuces.

The first mentioned of these, also known as tall lettuce and horseweed, is a towering plant, sometimes reaching over ten feet in height. It is an inhabitant of thickets, meadows, clearings, edges of fields and fencerows, and other damp, low-lying places. Its range extends from the Canadian Maritime Provinces to the West Coast and south from parts of the Gulf Coast to New Mexico.

A biennial, the common wild lettuce has a basal rosette of leaves arising from a taproot. These leaves are quite large, sometimes reaching a

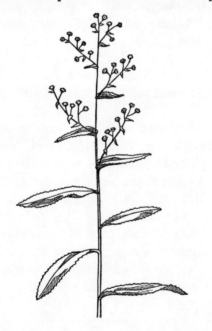

foot in length. Smaller leaves occur on the stem. The plant's lower leaves vary greatly in shape. They are irregularly lobed and toothed, and resemble giant dandelion leaves. Green on the upper surface and whitish underneath, the h a i r l e s s leaves arise directly from the stem, without leaf stalks. The upper leaves on the flowering stem are generally lance-shaped and not lobed and toothed.

The flower stalk, which emerges from the middle of the rosette, is branched near the top. The plant bears numerous composite yellow flowers, each about one-half inch high. The flower heads are sometimes shaped like an urn.

Wild lettuce bears a milky, white juice common to the other members of the genus *Lactuca.* This juice was used by the Indians in past decades to alleviate the itchy rash caused by contact with poison ivy. It was also used as a sedative and an inducer of sleep, and has an odor and taste similar to opium. Modern medical science dictates that the latex of the plant is, at best, an extremely mild sedative. Yet the milky liquid is reputed to calm the nerves and relax one's whole system.

The prickly lettuce, also known as compass plant because the tips of the leaves point north or south, is a resident of cultivated fields and gardens, roadsides, and other waste places. The plant occurs all over Canada, except in Newfoundland, and far into the United States. An immigrant from Europe, it is generally a more common plant than the native wild lettuce, and is often considered a nuisance weed.

This annual or winter annual sports a stout stem arising from a tap-root and often bearing prickly projections nears its base. The leaves are alternate, mostly deeply lobed and toothed, and clasp the stem by their bases. The leaf margins are somewhat prickly, and the whitish vein running the length of the leaf on the underside bears a row of stiff, spine-like bristles. The flowering stalk is highly branched, with small leaves. The composite flower heads, which appear from midsummer to early fall, are small, numerous, and pale yellow in color.

Like the other wild lettuces, the prickly lettuce bears a milky white sap which reputedly has mild sedative properties. The milky latex of this and other wild lettuces was at one time collected and dried, and used as a diuretic and hypnotic, as well as to calm the nerves. It is also reported that some Indian women brewed a tea of prickly lettuce leaves and drank it following childbirth to stimulate the flowing of milk.

The young leaves of both the common and prickly wild lettuces have been enjoyed by Indians and white men for many generations. The leaves and stems are very tender when young, but tend to toughen with age. They make delightful salads, especially if a zesty dressing is used.

Our favorite wild lettuce salad involves carefully washing the young leaves and tearing them into pieces the size of a quarter. Toss the greens with about 6 slices of cooked, crumbled bacon, lots of finely chopped chives or spring onions, salt and pepper to taste, ½ teaspoon of sugar, 1 teaspoon of vinegar and enough sour cream to coat the greens lightly and evenly. Chunks of tomatoes and hard-boiled eggs provide a nice garnish.

A creamy "green goddess" type dressing is also good over a tossed salad made of wild lettuce, sweet onions, small cucumber slices, bacon bits, and mustard greens. Beat 1 egg with a tablespoon of anchovy paste until smooth. Then add salt and 2 tablespoons of mild vinegar. Next, beat in 1 cup of vegetable oil, a little at a time, until smooth. Stir in 1 tablespoon lemon juice, 1 teaspoon onion juice, ¼ teaspoon garlic juice, a dash of dried tarragon, and a few strips of chopped parsley and chives. Then blend until smooth and refrigerate.

Wild lettuce is also good when wilted. For this, wash and tear up enough lettuce for four to six persons. Fry 6 slices of side bacon, and remove the bacon to drain. To the fat in the bottom of the frying pan, add 2 tablespoons vinegar, an equal amount of water, 1 tablespoon sugar, 1½ teaspoons onion juice, and salt and pepper to taste. Stir and heat, then pour over the wild lettuce. Add the crumbled bacon strips, and toss until the greens wilt.

The list of things that can be done with wild lettuce is almost endless. It can be braised in butter or added to soups or stews. It can be mixed with other leafy greens in a salad dish. It combines with almost any meat, cheese, or seasoning to make a casserole. A friend of mine once even made a wild lettuce pie! You guessed it. It *was* different!

Chapter II

POTHERBS OF THE WILDS

The old adage about variety being the spice of life certainly applies to the vegetable menu. Carrots, turnips, and string beans can quickly dampen even the liveliest of spirits when they appear on the table night after night after night. And despite all the disguises that can be tried in the kitchen, a carrot on the plate is still a carrot.

Wild potherbs, collected from spring to fall or cultured in the basement during winter, provide the variety that makes mealtime something to look forward to. The flavors of many wild potherbs cannot be matched by their cultivated counterparts. Others are similar to domestic favorites, but still provide delightful substitutes. And as if this were not enough, wild potherbs boast higher vitamin and mineral contents than cultivated vegetables, and, of course, are somewhat easier on the pocketbook.

Some care must be taken in preparing wild potherbs for the table. Although all of the plants in the previous chapter, Salads from Nature, can be eaten either raw or cooked, this is not the case for the potherbs. Many have to be cooked before they become edible. For example, the cowslip or marsh marigold contains a toxic substance when raw, but this material breaks down and disappears quickly with cooking. The other big advantage of cooking is that with a little imagination, the variations of recipes are almost endless.

MILKWEED
Asclepias spp.

My first introduction to wild edibles was in the form of boiled milkweed buds. I was quite young then and don't remember the taste too distinctly, but it must have been pretty good because milkweeds have been tops in my books ever since. They appear quite often on our dinner table, from early spring when we delight in the boiled shoots to late summer when we dine on the young seed pods.

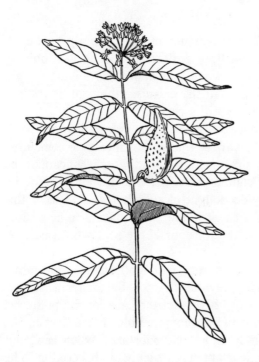

There are about a dozen species of edible milkweeds on this continent, but *Asclepias syriaca*, the common milkweed, is the most widely harvested. The common milkweed, or silkweed as it is also appropriately called, inhabits thickets, roadsides, fencerows, and dry fields, from New Brunswick to Saskatchewan, and south to Georgia, Tennessee, Iowa and Kansas.

Milkweed is one wild edible that should be harvested with a little caution, to ensure that only the right weed is collected. Two characteristics distinguish the milkweed readily; the silky, parachute-like seeds that tumble from giant seed pods and float lazily on the autumn breeze, and the milky latex that exudes when the stem is bruised or broken. The seeds are not around when you are harvesting young shoots in the spring, so here is where the milky sap comes in. It is important *not* to identify the plant solely on the basis of the latex it exudes. Some other species, notably the spurges and the dogbanes, are also latex bearers. However, they are not quite as agreeable to the digestive system.

The milkweeds are all unbranching and have stout stems growing from underground rhizomes. Their leaves are borne on short, sturdy petioles, in pairs, and on opposite sides of the stem. The oblong-shaped leaves are generally four to eight inches in length, with a prominent midrib running down the middle. A showy display of tiny, fragrant flowers appears in summer. These vary in color from near white to greenish-lilac to near purple. Each blossom is composed of five tubular hoods which eventually give rise to firm green seed pods, covered with wart-like bumps. At maturity, the pods split, releasing the silken seeds.

As a native North American, the various milkweeds were used by Indian tribes as medicines. The milky sap of one species was used by the Rappahannocks as a cure for ringworm. Other Indian tribes are known to have used "teas" made from the plant as contraceptives. There is even some indication that the French in eastern Canada may have made "sugar" from milkweed flowers.

The milkweed is one plant that should be utilized solely as a cooked vegetable. Actually, the appearance and bitter taste of the latex doesn't make the plant very appetizing as far as salads go. Generally the raw shoots of milkweed are regarded as toxic, or at least disagreeable. However, chewing gum can be made from the latex after it has dried, although I am sure it tastes better cooked.

Young, boiled sprouts are the first milkweed dish of the year. In early spring, when the shoots are about nine to ten inches long and the leaves have not yet unfurled, we cook up our "milkweed asparagus". We remove the fuzz on the shoot simply by rubbing it off. Then we cover the young shoots with boiling water and boil gently for about 2 to 3 minutes. After that we change the water and simmer until tender. Smother with butter, salt and pepper to taste, and a dash of fifty-fifty mixture of grated Parmesan cheese and bread crumbs. I should note here that some prefer to go through three or four water changes before eating, to lessen the bitter taste associated with the milky latex. We like a bit of the sharp, bitey taste of the milk, ourselves.

Milkweed flower buds are another delicacy, and may be a bit safer for the tyro because they cannot be mistaken in identity. We collect them when they are quite young, long before the blossoms have opened, and serve them in two ways — as a potherb and in salads — but *both* cooked. As a vegetable with firm, fresh, lake trout fillets, they are tops. Simply boil the buds briefly in two changes of water, smother with butter, and salt and pepper to taste. For salads, boil the buds beforehand, cool, and toss in with small bits of greens, mild radishes, and spring bunching onions. With a mild Italian dressing, they lend a unique flavor to the salad.

If you like okra, you are bound to like young seed pods of milkweed, harvested when still firm and fairly small. If the pods seem at all elastic to the touch, you have come too late. The seeds are too far developed for good eating. Again, the pods must go through several changes of boiling water before eating. A bit of baking soda in the water will also help to break down the somewhat tough fibers of the pod. Young seed pods are good served just with butter, or with butter and a touch of sour cream. One of our favorites is to stew the pre-boiled pods with fresh tomatoes, salt, pepper, a dash of thyme and a sprig of dill.

BURDOCK
Arctium spp.

The name burdock immediately conjures up an image of picking burs off sweaters and trousers and combing burs out of dogs' coats. Even the genus name *Arctium* is said to come from the word *arctos*, meaning bear. Alas the image connected with this plant is frequently an accurate one, but the Old World biennial weed known as burdock, burdane, and clotbur has such tasty roots that it is well worth fighting the burs to get the goodies.

Several species of burdock exist in North America, growing heavily in waste places and in dry and brushy areas. Fencerows, roadsides, and areas around old barns are favored habitats. Some burdocks like calcareous soil; most aren't terribly fussy.

The great burdock, *Arctium lappa,* is found from Quebec to the

Great Lakes area, touching the Canadian Maritime Provinces, and south into New England. The smaller or common burdock, *Arctium minus,* is widely spread from Newfoundland to British Columbia, south to New England, West Virginia and Virginia, and west through the midwestern states and into California.

Burdock was introduced into North America via the explorers and early settling colonists from Europe. It originated farther east, and has been prized for many years in Japan and other parts of the Orient as a vegetable.

Burdock is extremely easy to recognize, even for the tyro. The plant is a biennial and flowers in its second year. The maturing flowers produce the seed pods that are regarded as nothing but a nuisance by man and beast alike. The stem of the plant is long, frequently reaching six to seven feet in height. The large leaves are rough in texture, and sport a purplish color with prominent veins. The composite flowers that give way to the numerous seed pods are also purplish, and usually appear between July and October.

But what does one eat of the ungainly and pesky plant? The roots, the leaves, and the young flowering stocks can all be eaten. The roots were the oriental favorites, and come somewhere between radishes and parsnips in flavor and texture. They are quite strong, and some cooks prefer to boil them twice, changing the water between cookings, to alleviate the strength of the flavor and soften the tough fibers. We like ours in stews and in Indian curries, using them for flavoring. We have had some luck in freezing them too, so they last all winter. Only roots of first-year plants, before the flowering process is initiated, are recommended.

The leaves of the burdock are not the favorites of many, but they make nice salad greens when combined with other wild edibles or with raw spinach and a hefty blue-cheese dressing. Leaves for salads must be very young and tender. Young leaves are also good in soups or boiled with salt, smothered in butter, and sprinkled with Parmesan cheese.

Young flowering stocks of burdock, harvested before the flower buds have fully appeared, are also good. The outer covering of the stem must be removed completely before cooking, and it is a somewhat time-consuming project. As you are working, it is wise to submerge your finished or pot-ready stocks in water with a little lemon juice or vinegar added to keep them from turning brown. Then cook them in two boilings of water, cover with butter and a touch of salt, and you have a great vegetable. The young stocks can also be used, once boiled, in Chinese wok-fried vegetable and meat dishes.

MARSH MARIGOLD

Caltha palustris

The marsh marigold, or American cowslip as it is more commonly known, is one plant that should be collected with great care, in order to avoid mixing it with less agreeable or even poisonous species that grow

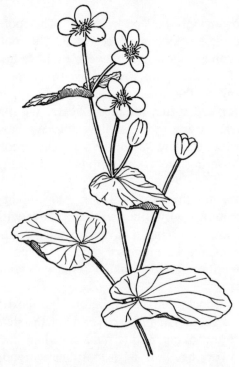

in the same habitat. With a little care and practice, the marsh marigold is easily distinguished from its poisonous neighbors. The plant proliferates in swamps, wet meadows and wet woodlands in the early spring of the year from Labrador, including Newfoundland, to Alaska, and south to Tennessee, South C a r o l i n a, Iowa, and Nebraska.

Cowslips grow between one and two feet in height, and are easily distinguished when their bright orange-yellow, buttercup-like flowers appear in late spring and early summer. However, the best eating occurs earlier, before the flowers have opened, and here is where care in identification takes place. The stems of the marsh marigold are fleshy, and the leaves at the bottom of the plant grow on leaf stalks that sport a papery, sheathing stipule at the base. The young leaves near the top of the plant rarely have significant leaf stalks. The leaves themselves are slightly rounded, heart-shaped, or kidney-shaped. They are a crisp, shiny green, and have scalloped leaf margins.

Two poisonous plants often occur in close proximity to the marsh marigold. These are the water hemlocks and the white hellebore, both of which are described in the chapter on poisonous plants. Neither really resembles the cowslip at all, but since they grow in the same area at the same time, the warning is felt necessary.

A second warning with regard to marsh marigolds must also be observed. The raw leaves of the plant contain a toxic substance known as helleborin. This substance is destroyed during any cooking process, thus marsh marigolds must be eaten only as a cooked vegetable and *not* used as a salad green. In fact, when you are out collecting in the spring of the year, it is a good idea to keep your cowslips in a plastic bag separate from your other greens. Then you are certain they won't get mixed in with the dinner salad by mistake.

New Englanders and Pennsylvanians have long regarded marsh marigolds as one of the true wild delicacies of spring. They are most commonly eaten, minus the tough, mucilaginous bases of the stipules, after boiling in two changes of salted water for an hour. Then they are drained and smothered with butter. We enjoy them this way, but we also like them with a cheddar cheese sauce.

Young, closed flower buds can also be used to make a rather interesting pickle, flavored somewhat like capers. These are made by boiling in two changes of salted water for about three hours, then simmering in vinegar, and canning in small jars. We made up a batch a few years ago and used them to decorate hors d'oeuvres. To say the least, they made interesting cocktail **conversation** pieces.

PURSLANE

Portulaca oleracea

For a plant of such ancient and honorable lineage, the common purslane is regarded by most modern-generation North Americans as one of the most pesky of garden weeds. Nothing short of complete removal — roots and all — will eliminate the plant from the suburbanite's landscaping efforts. If the weed eradicator digs a purslane plant up and simply throws the remains onto an old soil heap or another part of the garden, he is in for a big surprise. The succulency of the plant will allow it to remain very much alive until it comes in contact with the water and soil it needs to set up a new growth of roots. The purslane is a real example of the philosophy "never say die."

Purslane is a native of India and Persia. For over two thousand years, it has been appreciated in that part of the world for the culinary delight that it is. Moving on to Europe, it was quickly adopted as a choice vegetable. Purslane immigrated to America with the first settlers, and was even adopted by the southwestern Indians, who used the ground seeds as a breadstuff and meal. Today, only a few seek it out for table use.

The purslane is a creeping plant. Its prostrate form is often seen in dense mats of highly branched stems and thick, succulent leaves. The reddish-green or purplish stems are smooth, jointed, and very fleshy. From them arise narrow, wedge-shaped leaves, also reddish-green. The leaves occur in almost opposite pairs, and grow up to two inches in length.

By far the prettiest part of the purslane, or pusley as it is sometimes called, is its diminutive yellow flowers. They are somewhat inconspicuous at first glance because they grow directly in the forks where the stems branch, and because they show their golden faces only on bright, sunny mornings. But when they do let themselves be seen, they are a photographer's delight. Even some of the ornamental species of *Portulaca* capitalize on the beauty of the flowers for their appeal as bedding plants.

Another interesting part of the plant is its seed pods. Each seed pod has a little hat or cap that can be removed to reveal an enormous number of tiny seeds. However, by the time the seeds are around, the prime table life of the plant has passed. Purslane should be eaten early in the summer, when the plant is young.

Though sometimes restricted to specific localities, purslane generally grows wherever light soils occur. It is common in gardens, cultivated fields, and sandy waste areas. As far as distribution is concerned, the plant proliferates throughout the warmer, southern regions of Canada, most of the United States, and into Mexico where it is sold in the marketplaces.

When collecting purslane from your favorite garden patch, remember to pluck off only the leafy tips. If the remainder of the plant is left intact, it will go on to produce new leaves, and provide you with a continuous supply of greens from late summer to mid-fall.

Because purslane grows so close to the ground, the tips require a good soaking and rinsing before they get to the salad bowl. They can then be cut up or left whole, and used in any salad recipe.

As a cooked vegetable, purslane tips also hold their own. However, they are a bit mucilaginous in texture because of their succulence. This can be a bit of a blessing, because it means that you don't need a bushel of greens to make dinner for two. Very little of the purslane's bulk is lost through cooking.

For those who object to the mucilaginous texture, the problem can be solved by deep-frying. Simply wash the purslane tips, shake in seasoned flour, dip in beaten eggs, and roll in bread crumbs. Then deep-fry until tender. This takes care of the mucilage. We also use purslane in soups and stews, where it really lends a helpful hand in thickening.

One of our favorite cocktail-hour hors d'oeuvres is an Indian curry concoction, made from purslane tips. The tips are sprinkled with a little salt, after thorough washing, and allowed to sit for ½ hour. After this, they are individually coated in a batter made by combining 1¼ cups of chickpea flour (regular flour will do if necessary) ¼ teaspoon of ground cayenne, 1½ teaspoons of ground cumin seed, ½ teaspoon of ground coriander seed, and ¾ cup cold water mixed slowly with the dry ingredients. The batter-coated pieces are then deep-fried in hot oil for 8 to 10 minutes, or until golden brown, and then drained on paper towels. We serve them immediately with hot, spicy chutney, dried currants, and cashews. Our guests always manage to clean out the hors d'oeuvres plates.

Purslane is also a good candidate for pickling, and can be substituted in any pickle recipe calling for cucumbers. It remains crisp and flavorful as a pickle for many months if properly canned.

PLANTAIN

Plantago spp.

When I was a kid, the common plantain was a frequent resident of cracks between sidewalk stones where we played hopscotch, trampled corners of the schoolyard where nothing else could survive the ravages of a half thousand grade schoolers, and my parents' front lawn. Today, with the wider spread use of herbicides on lawns and sidewalk areas, it doesn't seem to be as noticeable, at least not in our part of the country.

Plantain is a short plant—completely stemless. The leaves, which are elliptical in shape, very broad, and have pronounced ribs running almost parallel down their length, arise directly from the root in a rosette. People who do harbor plaintain in their front lawns rarely see the whole leaves, because they don't have much time to grow between weekly attacks of the lawn mower, and few are familiar with the plantain flower. The plantain has many individual, small flowers, arising in a loose spike formation from the rosette of leaves. The greenish flower spikes rise quite high above the leaves, on long naked stalks.

The common plantain, *Plantago major*, originally from Europe, has a variety of forms, differing slightly in size and shape of the leaves and flower spikes. Basically they are all plants of roadsides, waste areas, and cultivated lands, including lawns.

The seaside plantains, *Plantago oliganthos* and *Plantago juncoides* are slightly different from the common species, mainly in leaf shape. They are deep-rooted perennials with long, narrow, fleshy leaves. The

leaves are generally erect rather than lying close to the ground, but also have quite prominent veins.

Known by the name of goosetongue, sea plantain is avidly collected by fishermen of the eastern seacoast for use as a cooked vegetable. Sea plantain is typical of saltwater marshes and brackish or saline shorelines, including gravelly or rocky saltwater areas. The plants range from Newfoundland and Labrador to Quebec and Nova Scotia, south to New Jersey, in the Manitoba marshes, and along the west coast from Alaska south to California.

Plantain leaves are very rich in vitamins A and C, and will remain so if the life is not cooked out of them. Just a small amount of water is needed, and a short time on the stove. When they are done, drain, salt and pepper to taste, and smother with butter. We like them with a sprinkle of Parmesan cheese on top, too. Their mildly bitter taste is best enjoyed if the greens are still slightly crisp when they reach the table.

As a salad green, the young leaves are very good. The older ones tend to be a bit stringy, due to the parallel veins. If the bitter taste bothers you unduly, try mixing plantain greens, ripped into small pieces, with other greens and your favorite salad dressing.

The Indians had many uses for plantain. The Shoshoni and other tribes used a heated, wet dressing of plantain leaves on wounds. Early settlers also used this dressing for cuts and bruises. Because of its antiseptic qualities, plantain was used for snake and insect bites, and to combat mouth infections. The Shoshoni also used poultices of plantain leaves to relieve rheumatism.

SCOTCH LOVAGE

Ligusticum scothicum

Colonists from the Hebrides and other maritime parts of Scotland were only too happy to find one of their favorite greens growing in abundance along the cold saline shores of Canada and the northern United States. Scotch lovage, one of nature's wild celeries, does not, however, enjoy the same popularity with Americans as it does with Scotsmen. This is a pity, because the celeryish taste can be delightful if a little care is taken in harvesting and preparing this herb for the table.

Scotch lovage looks something like domestic celery. It is a stocky, stout plant, with long leaf stalks arising directly from a perennial root. Each leaf stalk is crimson or purplish at its sheathing base, and is topped with three forks, each bearing one leaflet. The oval-shaped leaflets are a glossy green, and have coarsely toothed margins. The root of the Scotch lovage is heavy, thick, and deep.

As a member of the parsley family, the flower clusters of Scotch lovage are umbrella-shaped. The relatively flat-topped umbrellas are supported on thin, arching flower stalks, and are composed of many tiny white flowers. In fall these flowers give way to oblong-shaped dry fruits, nearly a half inch in length and pale brown in color.

Scotch lovage is a resident of ledgey and gravelly areas, windswept and rocky coastlines, sandy seashores, and saline marshes. In the east, it grows on the coast of Greenland, and from Labrador south to New York. It also grows in the temperate and cold regions of the West Coast.

There are as many recipes in the world for Scotch lovage—both raw and cooked — as there are for celery. Indeed the two are quite interchangeable, although Scotch lovage has a somewhat blander taste (or at least we think so). However, people who are taste sensitive claim that the wild celery is strong, and prefer to boil it in two changes of water before proceeding with any further culinary treatment. But we have a cold lovage chutney that we serve with Indian curries, and we have never had any complaints about the celery-like taste being too strong.

For our lovage chutney, we collect young leaf stalks with their leaflets, and wash them thoroughly in fresh, cold water. Because of their seaside habitat, they are often in need of a good bath before they present themselves at the table. Peel the stalks, and finely chop them with the leaflets. Take about 2 cups of the chopped Scotch lovage greens and mix them with 5 teaspoons of confectioner's sugar, 1 teaspoon of salt, ½ teaspoon of dry mustard, ½ teaspoon of black mustard seeds, and ⅓ cup of mild vinegar. After thorough mixing, let the chutney stand for a couple of hours in the refrigerator and pour off the liquid before serving. It is important to keep the concoction cold and slightly moist until just before serving, in order to retain the crispness of the diced greens.

Scotch lovage is also good as a potherb. There are several members of our family who regularly turn their noses up at braised celery, but all of them relish braised Scotch lovage, when we can get to the seashore to collect it. Perhaps it is partly the sea air that makes it taste so good, but whatever it is, braised lovage is a dish that calls for seconds. The dish is a bit time-consuming, but quite easy to prepare. First, wash and peel the young lovage stalks, minus leaves, and cut them into convenient lengths. Then pop them into a non-stick frying pan with a teaspoon of butter and just enough liquid to keep them from burning. Water is a suitable liquid, but chicken or any game bird stock is better. Now simmer until tender. Remove the Scotch lovage and arrange the stalks in a shallow baking dish—one that can be brought to the table for serving. To the lovage juices remaining in the frying pan, add ¼ cup of cream (at room temperature) thickened slightly with ½ teaspoon of flour and salt and pepper to taste. Mix thoroughly and bring slowly to a boil, stirring constantly. When the liquid is bubbling and thickened, pour it over the lovage stalks. Now, sprinkle the whole dish liberally with grated sharp cheddar cheese, and pop under the broiler for a few minutes until the cheese melts. We have always found this a favorite with firm, thick, fish steaks.

ANGELICA

Angelica atropurpurea and *Coelopleurum lucidum*

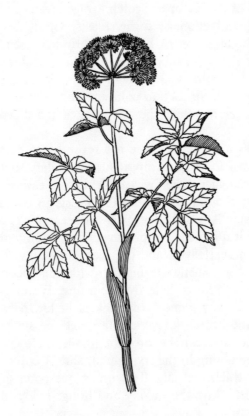

Now here is where some confusion comes in with names. I think a couple of people must have had some grudges or bones to pick when naming the angelicas, because to anyone but a taxonomist the whole matter of the wild celeries is more than a bit confusing.

The plants we are talking about here are both members of the parsley family. What we know as "wild celery" is *Angelica atropurpurea*, probably more correctly named purple angelica. This is a tall and stout plant with smooth, purple stems. From the stem emerges many leaf stalks, each with a broad, heavy sheath at its base. The coarsely toothed, slightly oval-shaped leaflets occur in groups of three atop the leaf stalks. Because the leaf stalks do not arise directly from the root, the purple angelica does not look as much like domestic celery as the Scotch lovage does.

The umbrella-shaped flower heads of the purple angelica are made up of tiny, greenish-white flowers. The whole umbrella structure looks vaguely like a greenish-white globe, about three to six inches in diameter. The purple angelica often grows from four to six feet in height, and the plant has a rather pleasant, aromatic odor about it.

Purple angelica is a plant of rich thickets, swamps, marshes and other bottomlands from southern Labrador and Newfoundland to Wisconsin, the Midwest, and east to New England and the Atlantic seaboard. The plants flower from late May to September, depending on latitude, and fruit from July to October.

The seacoast angelica is not a member of the genus *Angelica* at all, (so some authorities say). However, it is very closely related. The seacoast variety of wild celery is a coarse plant with large leaves, each bearing a prominently inflated sheath at the base. Again, the leaflets are arranged in groups of threes. The leaflets are coarsely toothed, more or less oval shaped, and green on both top and bottom. The stems are coarse, green in color, and adorned with sticky spots.

Again the umbrella-shaped flower head prevails. The individual flowers are small and white, and give rise to small, oblong, ribbed dry fruits. The whole plant grows from one to four feet in height.

The seacoast angelica is a plant of rocky and gravelly coastlines, ledgey seashores, and coastal thickets. Its range extends from southern Greenland and Labrador to New York, including Quebec and the lower St. Lawrence River valley.

Both of the wild celeries are considerably stronger in taste than Scotch lovage. (If you are simply going to boil them and smother them with butter, perhaps you had best change the water once). We clean and peel the young angelica stems and leaf stalks, cut them into 3-inch lengths and barely cover them in a saucepan with lightly salted water. Then we bring them to a boil and simmer for 20 minutes or until the stalks are tender. Next we drain them and smother with butter.

We have another favorite wild celery sauce. In a small, heavy saucepan over low heat, melt 2 tablespoons of butter and stir in 2 tablespoons of flour. When the mixture is well combined, slowly stir in one cup of whole milk, stirring constantly to keep the sauce smooth. Slowly bring to a boil, stirring all the time, and add salt and pepper to taste. Now add ¼ cup of toasted almond slivers and pour over the cooked wild celery arranged on a serving platter.

Cream of wild celery soup is always a delight, and very easy to prepare. You need about 2 cups of finely chopped angelica leaflets and leaf stalks in your blender jar. Add one small onion finely chopped and 2½ cups of chicken broth. Now blend until smooth—a few seconds should do it. Add 1 cup of whole milk and ¾ cup of cereal cream (or milk), salt and pepper to taste, and bring to a slow boil. Serve hot, with fresh rolls and butter. A little flour will be necessary if you want a thicker soup.

CLOVER

Trifolium spp.

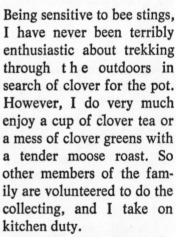

Being sensitive to bee stings, I have never been terribly enthusiastic about trekking through t h e outdoors in search of clover for the pot. However, I do very much enjoy a cup of clover tea or a mess of clover greens with a tender moose roast. So other members of the family are volunteered to do the collecting, and I take on kitchen duty.

The clovers are known by young and old alike for their fragrant, honey-laden blossoms and their dainty, three-leafed stems. Many a child has spent hours in search of the elusive four-leafed clo-

Red Clover   *Alsike Clover*

ver but alas, except in the west where five- and six-leafed specimens occur, the three-leafed plants predominate. Indeed, even the genus name *Trifolium* is a product of the words *tres* meaning three and *folium* meaning leaf.

Well in excess of seventy species of clover grow in the temperate regions of North America, frequenting meadows, pastures, open fields, gardens, roadsides, and even lawns, where it is often none too popular with the homeowner. We have never been overly bothered with the few patches of this legume that inhabit our lawn. The plant sustains its own as a nitrogen fertilizer, and because it is so deeply rooted, it remains green even when the grass is longing for a good rainfall. However, to some people it is a nuisance because it stains children's clothing and leaves bare spots in the landscaping once the leaves have been killed off by the first frost. In Vermont, the red clover is thought of highly enough to have the distinction of being the state flower.

A member of the pulse or pea family, clover is essentially spreading in habit. The stems bear few to many leaves, each sporting a stipule at the base of the petiole, and three finely toothed leaflets. The flowers vary in color from white and yellow to a purplish-red. Avidly pollinated by bees, the first clovers introduced to Australia failed to reproduce until the bees were imported as well. The flowers have a sweet odor, and a honey-like taste when eaten raw. Clover has long been used as a livestock forage, and is sought by all manner of wild creatures — birds, as well as mammals.

Clover was used by a number of Indian tribes as a foodstuff. Several tribes gathered the leaves, before flowering, and consumed them fresh or cooked, alone or with other edible greens. In some areas, clover was even helped along in its growth by irrigation. Some Indian tribes held yearly spring rituals to welcome the new crop. In Scotland and other parts of the British Isles, a nutritious and wholesome bread was made from dried clover flowers and seeds, and was important during times of food shortages.

One interesting point about clover is that too much of it is known to cause bloating, in both cattle and humans. Dipping clover in salt water was thought by some Indian tribes to solve the problem.

We best enjoy clover as a tea. When the flowers become ripe and dry, either in fall or during a dry spell earlier in the year, we collect them and further sun-dry them on the back porch. We then crush them with a rolling pin, or buzz small amounts around in the blender for a few moments, and store the dried flowers in an airtight container. When we are looking for a different and refreshing hot drink, we steep the dried clover in boiling water in a tea pot, about ¾ to 1 teaspoon per cup. The "leaves" make interesting reading once you have finished your tea.

For use as a potherb, gather a good batch of young clover leaves — about 8 or so cups. Then fry a small chopped onion, until translucent, in a non-stick frying pan with a little butter. Add ½ teaspoon of celery seed, a little salt, and stir. Then add the clover greens, cover, and simmer until the leaves are limp and tender. Next, remove the cover and toss the leaves and onions, mixing thoroughly. Serve immediately as a deeply green nutritious vegetable.

NETTLE
Urtica spp.

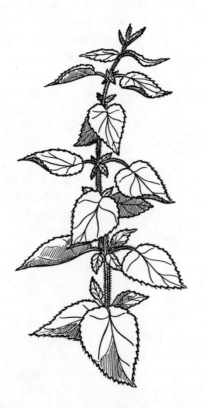

To those who don't know about all the good qualities of the nettles, the plants appear as one of the scourges of the earth. They invade roadsides and r u b b i s h heaps, colonize playgrounds, and occupy front lawns. Their sharp spines prick and sting anyone who tries to remove them by brute force, and an accidental brushing against a nettle will produce a stinging of the skin followed by a reddish rash.

However, the nettles do have their good side. Aside from being very rich in vitamins A and C, they have a very high protein content.

In some areas nettles are cut and dried, and fed to livestock when mixed with regular rations. Although no bovine will look twice at a nettle when it's growing in the ground, all livestock, as well as poultry, will avidly devour dried, chopped nettles as part of their diet and be healthier for it.

The nettles all make good eating, once you have learned how to harvest and cook them, avoiding the relentless "stingers." Indeed they are an excellent emergency food when boiled because they are readily available and easily recognized. In addition, the stems of the older plants provide a strong, fibrous material that can be used as rope or fishing line, and perform other useful functions in the wilderness.

The first line of defence in collecting nettles for the table is a stout pair of gloves. Although heavy cotton gardening gloves are sometimes satisfactory, leather or cotton gloves coated with a plastic material are much better. The strong stingers of the nettles can quite easily penetrate regular cotton gloves. Long gloves are better than short ones, as the harvester's wrists often become attacked by the nettles when reaching down to cut the plant off at its base. A knife or a pair of pruning shears are needed to cut the stem of the nettle. When collecting, it is best to cut the plant and immediately toss it into a basket or other carrying device for easy transportation.

If you should become stung by a nettle, the sting and redness will pass away fairly quickly. However, if it bothers you, a dab of rubbing alcohol will alleviate the problem. Some say that the juice of dock leaves *(Rumex crispus)* applied to the stings will relieve the discomfort. The Indians are known to have rubbed the irritated area with the dry, rusty-colored material that covers young fiddleheads.

Both annual and perennial nettles occur in North America. The common stinging nettle, *Urtica dioica,* is an erect plant with opposite, coarsely toothed leaves. These leaves are somewhat oblong or oval-shaped, and are covered with hairs and with stiff, stinging bristles. The leaf petioles and stems also bear these bristles, which are faintly swollen at the base and contain formic acid. At the top of the plant, between the leaves and the stem, appear slender, forking clusters of flowers. These are tiny and greenish in color, and are easily overlooked.

Several species of nettles are native to North America; others have been introduced from Europe and Asia. The nettles abound in waste places, on roadsides, in rich thickets, near garbage dumps and in gardens and lawns throughout the temperate regions of this continent.

As with collecting in the field, a little care is required in handling nettles in the kitchen. Only the very young plants, before they have reached ten inches in height, are good to eat. The older plants are tough and bear stonelike cells in the leaves that render them unpalatable. However, even the young plants sting fiercely when handled. So keep your gloves on until you get the plant safely soaking in cold water in the kitchen sink.

After a thorough rinsing, the plants can be lifted with tongs into a non-stick saucepan. Only a small amount of water should be used with the greens, and the leaves should be simmered just until tender. Cooking completely negates the stinging properties of the plants. When the greens are tender, drain the liquid and use it in soups. Then salt and pepper the greens and smother them in butter. They make a delicious potherb.

If you wish to chop the nettle greens before cooking, they can be held with tongs in one hand and cut up with kitchen shears held in the other hand. For soups and purées, the greens can be cooked and then passed through a sieve or buzzed around in a blender. Nettles can be used in any recipe calling for spinach or beet greens.

We use nettles to make a filling for ravioli. Making the pasta is a lot of work, but the nettle filling makes it all worthwhile. For the filling, we cook enough nettles to give us about 6 cups of chopped, cooked greens. To this, in a large mixing pan, add salt and pepper to taste, 1 cup of bread crumbs, 1½ cups of grated Parmesan cheese, ½ cup of grated Romano cheese, a dash of oregano, and 4 eggs. Mix this thoroughly, and allow it to stand, covered, in the refrigerator while kneading and rolling the pasta. Then stuff the ravioli (any good cookbook will give you the details of how to go about this) and seal them with beaten eggs. Place the ravioli on a cookie sheet covered with waxed paper, alternating between layers of paper and ravioli. Refrigerate overnight to age the pasta and ensure that the small pies remain tightly sealed. The following day boil the ravioli in slightly salted water until the pasta is tender and the filling hot. Drain and cover with your favorite Italian-style meat or mushroom sauce. This is a traditional birthday dish for one member of the family who was fortunate enough to come into the world when the nettles were young and harvestable.

CHICKWEED
Stellaria media

As a none too enthusiastic student in a course entitled "Introduction to Taxonomic Botany," I can remember setting out one fine Saturday morning to search the wilds of the Montreal city parks for plant materials that I could use to practise preparing mounted herbarium speci-

mens. In the first green area I came across, a small park a few blocks from where my parents presently live, grew mounds upon mounds of chickweed. "Well, nothing interesting here," I thought, "I can get chickweed right out of our own front lawn!" But I collected some anyway, from this park and from two others close by. Back home, I carefully pressed my chickweed specimens to dry them, and discovered that among my collectees were vegetative plants, plants with flower buds, plants in full bloom, and plants forming fruits — all collected within a relatively small area and all on

the same day. When our first mounted herbarium sheets were due the following week, mine was a composite of a good portion of the flowering and fruiting cycle of *Stellaria media,* the common chickweed. Though it may only have been a lowly weed, it earned me an "A," and earned itself a chance to be tried out on our dining room table.

Chickweed is a very common plant in North America, and indeed throughout the rest of the world. Presumably inheriting its name from the fact that the leaves and seeds are enjoyed by young game birds and domestic chicks, the plant is a member of the pink family. It is common in damp woods and thickets, in waste places, and in cultivated fields and gardens. It also happily invades lawns and is quite common in back yards and parks, except where spraying with the herbicide 2,4-D readily tolls its death knell.

The plant can be an annual, a winter annual, or a perennial, depending on where it grows. Chickweed flowers throughout most of the year in areas where the winters are not too severe. However, the flowers show their faces only on sunny mornings. The tiny, white blooms have five petals, each with two lobes, and produce somewhat oval-shaped capsules with tiny projections. These are the fruits that are relished by many wildlife species, mammals included.

The stems of chickweed are bright green, highly branched, and have slightly swollen nodes. From each node arises two, opposite, oval-shaped leaves with pointed tips. The stems are fragile and can rarely support themselves, hence the plant is usually spreading and does not often grow over a few inches high. Only when the plants occur in densely matted patches can the stems support one another sufficiently for the plants to attain any height. The leaves of chickweed are sometimes slightly hairy, while the stems sport a narrow band of whitish hairs running up one side of each of the many branches.

On the table, chickweed is fairly bland in flavor. It can be used in salads, but usually requires some sprucing up in the form of adding more pungent greens (mustard or watercress will do nicely), goodies such as crisp bacon bits, seasoned croutons or crumbled blue cheese, or a zesty salad dressing.

It can also be good in soups and stews, or as a potherb. Chickweed leaves require very little cooking, and should be quickly done with as little liquid as possible. Cooked briefly in salted water, drained, and smothered in butter, they can be quite good — but again too bland for our taste. On several occasions we have mixed chickweed with spinach or Swiss chard and used it as a boiled vegetable with butter and a sprinkling of grated cheese or bread crumbs.

One favorite Sunday morning breakfast back home incorporated chickweed greens, or at least it did after my success with the chickweed herbarium specimen. The dish was a traditional recipe in our household, but the greens made it so much the better. For six of us who were always hungry for brunch about eleven o'clock Sunday morning, we took 12 slices of cracked wheat bread and removed the crusts all around. (The crusts were saved for game bird stuffing for dinner.) Then with a pastry brush, we spread melted butter with a touch of garlic on both sides of the bread slices. We pressed the pliable slices into a large muffin tin, one slice per muffin slot, and made "cups" out of the bread.

Then we mixed a dozen eggs (we were all hungry college students home for the weekend!), 1 cup of milk, a drained can of mushroom bits, salt and pepper, 1 cup of finely shredded chickweed greens (packed), and 1 cup of shredded mustard leaves (from our cultivated patch in the back yard) in a very large bowl. When the ingredients were mixed thoroughly, we scrambled the egg mixture in a non-stick frying pan in butter, stirring the eggs as they cooked until they were soft and creamy.

In the meantime, the toast cups were baking in the oven until browned and heated through. The finished chickweed-and-mustard scrambled eggs were scooped into the toast cups and served with thick slices of ham, fresh corn bread, homemade preserves, and lots of coffee. A memorable breakfast that stemmed our student appetites — at least for a couple of hours!

ROSEROOT

Sedum rosea

Although not a member of the rose family, the roseroot does bear semblance to the roses in one aspect. If broken or bruised, the roots of the plant emit an odor comparable to that of a pot-pourri made of strongly scented rose petals. The plant is also known as stonecrop and rosewort, and in some areas is called scurvy grass because it, too, provided a source of vitamin C to the explorers and adventurers who were a part of this continent's rich heritage.

The roseroot is a familiar sight on the rocky coastlines of Labrador and Newfoundland. It is abundant in the arctic coastal regions as well, and stretches its reaches southward to Maine in the east and British Columbia in the west, and locally as far as North Carolina. In restricted locations, it even occurs as far inland as Vermont and Pennsylvania, although it is not generally an inland dweller. The roseroot is a resident

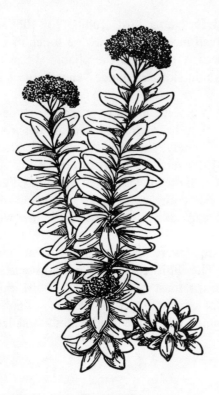

of rocky sea cliffs and ledges.

If there is any doubt in the mind of the wild harvester as to the correct identity of the roseroot, the telltale odor of a scraped root will confirm the identification. The thick, strong roots give rise to numerous stems up to ten inches in height. Fleshy, pinkish to whitish-green leaves crowd the stems. The succulent, oblong leaves bear both toothed and smooth margins and are quite pale in color. At the top of the stem appears a dense cluster of pale yellow, four-petalled flowers bearing pollen. The purplish seed-bearing flowers, also part of the cluster, produce reddish to deep purple seed pods resembling four- or five-pronged capsules.

Although you may not be too enthusiastic about eating great quantities of anything that smells like a florist's shop, the succulent leaves and stems of the plant are very mild in flavor and quite agreeable. They are tender when young, but toughen as the plant ages. By the time the seed capsules appear, the perennial is best left until next year for a harvest of greens.

Roseroot can be used both as a salad ingredient and a cooked vegetable. Even the roots can be mixed with other cooked greens to lend a somewhat unusual flavor. Simply boiled with salt, sprinkled with a bit of pepper, and smothered in butter, the leaves are quite tasty. In the very northern regions of the plant's range, on the tundra, roseroot is commonly eaten, as it is one of the few edible wild plants available where the growing season is so short.

Roseroot leaves are enjoyable stewed with fresh, peeled tomatoes, a little thyme and dill, and salt and pepper to taste.

ORACH

Atriplex spp.

Another member of the goosefoot family that is not terribly attractive to look at yet a treat on the dinner table is orach, sometimes known as saltbush because of the seacoast and other brackish water environment in which it is common.

This plant is known to have been used by the Navajos as a treatment for bee and wasp stings. The Indians made a pulpy mash of the plant and applied it to swollen bites. The Zuñis are also reported to have used the dried plant, mixed with saliva, as a remedy for ant bites. As a foodstuff, the stems and leaves of the plant were boiled with other greens and meats to make soups and stews. The seeds were also ground and used as flour.

Orach is quite easily recognized by those who seek it. Except for details of the flowers and fruits, it greatly resembles lamb's-quarters. It is characteristically different from lamb's-quarters and strawberry blite in that the pollen-bearing and seed-bearing flowers are borne on different plants. In addition, each fruit has two large sepals, somewhat resembling wings, arising from its base. The plants are highly branched, and bear the triangular-shaped leaves of the strawberry blite.

Saltbush is very common along both seacoasts of North America, mostly in the temperate zone. It also occurs inland in rich, open soil areas across most of the continent, as well as near dry lakes and in salt marshes.

As a relative of beet and spinach, the plant can be treated much like the latter for table use. However, because of its somewhat salty nature, care should be taken in rinsing it thoroughly before cooking and in using no added salt in the pot.

Young leaves picked in late spring or early summer make an excellent wild spinach soup, distinctly different from lamb's-quarters soup. Chop the thoroughly rinsed orach leaves and simmer them in water until tender. Then put a cup of the chopped, cooked leaves in the blender jar with 1½ cups of milk, 1 cup of chicken or game bird broth, ½ cup of cream, one small chopped onion, and a sprig or two of fresh parsley. Blend until smooth, then heat slowly until the soup reaches the boiling point. Be careful not to burn the creamy soup. When the soup is close to boiling, add a chicken bouillon cube and a dash of pepper. Then serve hot with crisp croutons floating on the top. A sprinkling of grated Parmesan cheese also adds a nice touch.

Orach also makes a good salad, but again with a minimum of salt in the dressing.

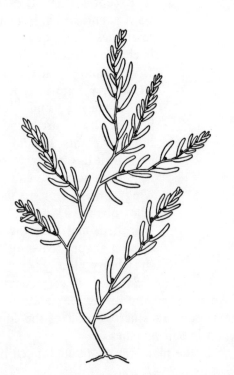

SEA-BLITE

Suaeda spp.

Because we have never been lovers of excessively salty foods, sea-blite has not been one of our favorite edible wild plants. However, we have eaten it on several occasions when visiting friends on the seashore, and taken in small quantities, its biting salty taste can be enjoyed. It is not a plant that is recommended as a steady diet for someone with high blood pressure, though.

The plant is somewhat similar in nature to its goosefoot family cousins. The alternate leaves are fleshy and rather cylindrical or linear in shape. The flowers arise directly from the axils of the leaves or leafy bracts.

Five species of sea-blite inhabit the coastal salt marshes and sandy sea-strands of the temperate regions of this continent. One of these also abounds in saline soils across the western plains. The Indians of the southeast and the west are reported to have eaten boiled sea-blite leaves as a green vegetable, and to have ground the seeds into flour.

We have tried boiled sea-blite leaves, but found that we had to soak them thoroughly before boiling and to change the water at least twice while cooking to reduce the salt content. Friends of ours mix the boiled leaves half-and-half with cooked Swiss chard, and this reduces the salty taste considerably.

GREEN AMARANTH
Amaranthus spp.

Close to a dozen amaranths occur in North America, and at least two of these are commonly known as green amaranth. Wild beet is another name for the plants, owing to their deep pink or red roots. As a close relative of the goosefoot family, the plants are also occasionally called pigweed.

The small, black seeds of the amaranths were long favored by a number of Indian tribes who parched the seeds and ground them into meal or flour. Reports also have it that some Indians even cultivated the plants.

The amaranths are quite similar to their goosefoot cousins in appearance. Only the inflorescences and the slight hairiness of the stems and leaves differ significantly. Some of the amaranths are quite tall, growing up to six feet. Their stems are rough, robust, and normally do not branch. The leaves arise from the stem on long leaf stalks. The leaves themselves are quite long, oval-shaped with pointed tips, and prominently veiny beneath.

The flowers of the green amaranth are not nearly as attention-getting as those of its close relative the purple amaranth, another common weed throughout the continent. Whereas the purple variety's flowers are a brilliant reddish-purple, those of the green variety are an insignificant green. The flowers are borne in loose clusters arising from the leaf axils. The long, branching flower spikes bear numerous, small, black seeds at maturity.

The green amaranth is not an unattractive plant to look at, and its flavor is subtle — almost delicate — yet not as bland as some other species. And aside from being high in vitamin content, as most fresh greens are, the green amaranth is chock full of minerals, especially iron.

The growing tips, young stems, and young leaves are the most tender parts of the plant. But even the flowers are good before they have become too old. The plant can be enjoyed in a salad or as a potherb, either by itself or in combination with more pungent herbs such as those belonging to the mustard family. We also like green amaranth soup, made in much the same way our other "wild spinach" soups are made.

Green amaranth is particularly enjoyable when prepared with sour cream. For this, start with about 8 to 10 cups of shredded leaves and stems, boiled in a little water until tender. Drain, and set aside for a few minutes. In the meantime, fry about ⅓ pound of side bacon, chopped into small pieces, in a non-stick pan. When the bacon bits are done, remove them from the pan and pour off the excess fat, leaving only a light surface coating of fat in the pan. Sprinkle 2 tablespoons of flour into the drippings and quickly stir until smooth. Then add 1 cup of commercial sour cream (or the homemade stuff if you are a purist!). When the sour cream mixture is hot, add the cooked bacon bits and salt and pepper to taste. Now you can pile the greens onto a serving platter and pour the hot sauce over the top, or toss the cooked greens right in with the sauce and heat thoroughly. This dish is delicious with firm, white lake trout fillets.

FIREWEED
Epilobium angustifolium

The fireweed gets its name from its persistent habit of popping up and colonizing burned-over areas, almost as soon as the smoke has died down. Over vast tracts of land blackened by flames, the quietly undulating waves of magenta flowers are like a soothing sea of tranquility.

Fireweed has long been a favorite wild edible of white men and redskins alike. Many Indian tribes feasted on the young shoots and stems of the plant, and even in Europe today, the plant is a favored vegetable.

A tall perennial plant, sometimes reaching seven to eight feet in height, fireweed occupies many new clearings, especially near logging operations, road clearings and damp ravines and ditches, as well as areas burned by forest fires. It grows in abundance throughout Canada, including the far north, and into Alaska. The range of the plant extends

as far south as Maryland and the Carolinas, Kansas, and California, and locally to Mexico.

The stems of fireweed arise from creeping rootstocks. Long, narrow, willow-like leaves give the plant its alternate name of great willow-herb. The most easily recognizable feature of the plant is its long spike of pinkish to magenta or purplish flowers. T h e s e loo e pyramids of flowers top the single stems, and mature to form long and narrow pods filled with tiny seeds. Each seed sports a tuft of white hairs that helps it "fly" to the place where it will begin another generation.

The tastiest portion of the fireweed is the young vegetative shoot, shortly after it emerges from the ground. Young shoots can be prepared like asparagus, although the flavor is not the same. The thoroughly washed shoots are good simmered in a non-stick pan in about ⅓ cup of butter and ⅓ cup of water until tender, with salt and pepper added to taste.

Young fireweed shoots are also delicious when cut into pieces, boiled in a little unsalted water, drained, arranged in a serving dish, and covered with a curried cream sauce. For the sauce, stir 2 tablespoons of flour into 2 tablespoons of melted butter over low heat. Then add ½ teaspoon onion salt, ½ teaspoon sweet paprika, 1 teaspoon prepared curry powder, and ¼ teaspoon ground ginger. Stir in one cup of cold milk or cereal cream, and keep stirring until smooth and heated through.

When the plants have become older, both the leaves and the stems can be eaten, but the stems should be peeled and cut into short sections before boiling. When drained, the greens can be covered with melted butter with a little lemon juice added.

KNOTWEED
Polygonum spp.

Depending on whose word you are taking, there are somewhere between thirty and fifty species of knotweed in North America. They are found all over the continent, and are easily recognized even by the tyro wild harvester. None of the species is reported as poisonous, so all members of this buckwheat relative make excellent emergency rations. One word of warning, though. The acid content of the knotweeds is fairly high, so it is best to either cook the plant before eating it or restrict yourself to small nibbles, in case you promote intestinal disturbances.

Of the many knotweeds, the Japanese knotweed, *Polygonum cuspidatum,* is perhaps the best known. Many generations ago it made its way to the New World from eastern Asia. It quickly established itself on this continent, and today is so widespread in some areas that it is considered to be a bothersome weed. The Japanese knotweed favors

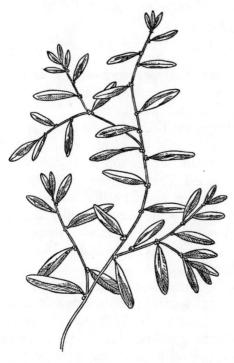

unattended cultivated fields and gardens, roadsides, shallow ditches, and other neglected bits of land. Its range stretches from Newfoundland across southern Canada to Minnesota, and south to North Carolina and Missouri.

The Japanese knotweed differs from the other members of the genus in having leaves that are sharply pointed. The plant is shrub-like in appearance, arising from heavy underground rhizomes. The upright stems of this perennial stretch up to eight feet in height. They are covered with a whitish substance or sometimes mottled, and bear prominently enlarged joints or nodes. Each node is surrounded with a thin, papery membrane. The somewhat rounded or oval-shaped leaves arise on stalks, and terminate in an abrupt point.

From the leaf axils grow branching flower stalks, each supporting numerous greenish-white flowers. Large, shiny seeds abound when the plant has matured.

Although many people eat the young leaves and roots of the Japanese knotweed, we prefer the young shoots. Usually the shoots can be recognized easily because last year's old stems are still standing close by. Once the old stems die, they dehydrate and form a bamboo-like material, rattling noisily if the wind disturbs them.

The young shoots should be harvested when they are under a foot in height, before the leaves begin to unfold. The shoots can then be washed and used as a substitute for asparagus, although the two differ greatly in flavor. Japanese knotweed can be simmered in a little water and will become fork-tender very quickly. Smothered with butter and a little salt and pepper to taste, it is a delicious addition to any spring dinner.

We have our own version of eggs benedict with knotweed, and it is a special Sunday brunch for everyone. For this hearty repast, halve and toast one English muffin for everyone. Onto the muffin halves, place a thin slice of ham that has been quickly sautéed in butter. Then pop a poached egg onto each half. Arrange two eggs per plate, and garnish on both sides with three or four shoots of knotweed that have been simmered until tender. Then cover with hollandaise sauce.

There are about as many recipes for hollandaise as there are seeds on a stem of knotweed, but our favorite is the old, traditional one, thinned a little for this particular dish. In the top of a double boiler, beat 3 egg yolks with a wooden spoon. Then add ¼ pound melted butter, 2 tablespoons lemon juice, 2 tablespoons hot water, salt and a touch of pepper to taste, and stir constantly while slowly heating. Beat until the sauce begins to thicken. Then add warm cream, 1 tablespoon at a time until the sauce is the consistency you like. Remember, though, that this sauce thickens considerably as it cools.

Japanese knotweed is not the only member of the genus that makes good eating. The giant knotweed or sachaline, *Polygonum sachalinense*, was long cultivated in Asia and Europe as a highly desirable potherb. Introduced to North America in the middle of the nineteenth century, it has spread rapidly through neglected gardens and wastelands. The young shoots and leaves of this tall knotweed taste good when simmered for a short time in a little salted water and then smothered in butter.

The young stems of the giant knotweed, when peeled and cooked, make an excellent rhubarb substitute, with a faint trace of a lemony taste. Stewed knotweed stems are delicious, with a distinct tartness of their own. To stew the stems, cup them into 1-inch pieces, add sugar in the proportion of about 1 part sugar to every 4 or 5 parts knotweed, and toss in a touch of cinnamon and nutmeg. Cover and simmer for about 15 minutes or until the stems are soft and well stewed. Taste and add more sugar if you feel the need for it. Served hot with vanilla ice cream melting over the top, it is hard to beat.

GOATSBEARD

Tragopogon spp.

The three species of *Trago-pogon* that exist in North America bear three more or less interchangeable common names. The first, goatsbeard, arises directly from translation of the Greek w o r d s *tragos* meaning goat and *pogon* meaning beard, and refers to the numerous long plume-like bristles that the fruiting heads bear. The name salsify is perhaps m o r e commonly known, one of the three species being the salsify that is cultivated on this continent and across the Atlantic for its tender young roots. The third name, oyster plant, aludes to the oyster-like flavor that some people attribute to the plant. I find the flavor to resemble parsnips much more than the shellfish for which it has been named.

The goatsbeards are all basically similar in appearance. They are fairly tall, biennial and sometimes perennial plants, with straight, fleshy taproots. In their first year, a broad basal rosette appears. The second year produces a tall flowering stem, sometimes branched, and abundantly adorned with broad, grass-like leaves. The base of the leaves clasps the stem, and the tip curls backwards away from the stem. The leaves, like those of the dandelion, exude a milky sap when broken or torn.

When flowering and fruiting time comes along, the three goatsbeards are most readily distinguishable. The cultivated salsify or oyster plant,

Tragopogon porrifolius, bears purple flowers and fruits that are less than one-half inch in length. The common goatsbeard, *Tragopogon pratensis,* sports yellow flowers, a flower stalk that is narrow right to its apex, and fruits about one inch in length. The larger goatsbeard, *Tragopogon dubius* (sometimes dubbed *T. major* because of its size), also has yellow flowers, but its flowering stem is thick near the summit and the fruits range from one to two inches in length.

As members of the Compositae, all of the goatsbeards have composite flowers. The flower heads arise singly on the long stalks, and the fruits bear a slender, terminal "beard." The plants flower from late spring throughout the summer, but sometimes close their blooms on very bright days.

The goatsbeards occur universally across most of southern Canada and the United States. Introduced from Europe, they inhabit roadsides, railway banks, rocky ditches, fields, abandoned pastures, old meadows, and numerous other waste places. The cultivated salsify has also escaped the farmer and gardener in many places.

Young goatsbeard leaves make a passable salad, but are better enjoyed as a potherb. Boiled in a little water or game bird broth, the greens are good simply covered with butter, sprinkled with a little lemon juice, and seasoned to taste with salt and pepper.

Salsify roots make even better eating than the leaves. The wild roots are smaller and thinner than the cultivated ones, so they take a bit more time and care in preparing for the table. However, they are worth it. The roots should be harvested when the plants are late into their first year or early into their second, before the flowers develop. After that the roots, like old parsnip roots, become very pithy and tough. They should be scraped of their outer covering and washed thoroughly before cooking. If you are preparing the roots ahead of cooking time, they should be kept in a little acidulated water to prevent them from turning brown.

We enjoy goatsbeard roots in several ways, but always partially cook them before incorporating any other recipe ingredients because they take some time to soften. Our pre-cooking is done either in a saucepan with a bit of water for about 30 minutes, or in a pressure cooker with less water for about 5 minutes.

Our candied salsify is made by cutting the cooked roots into inch-long pieces, placing them in a shallow baking dish and covering them with a mixture of melted butter and dark brown sugar. Then into the oven at 400 degrees for about 20 to 30 minutes, turning the roots a few times to ensure that they become thoroughly covered with the brown sugar syrup.

Several dishes can be made from thoroughly cooked, mashed goats-beard roots. One of them is salsify and cheese casserole, a dish that has long been popular in our house as a menu-mate for tender slices of braised moose. For this dish, we use roots that have been thoroughly boiled in salted water, drained, and mashed with milk until smooth. We then lightly butter a casserole dish and place half the mashed roots in the dish. Sprinkle with a little pepper and oregano, and then cover with a layer of grated cheddar cheese. Now add the remaining roots and another layer of spices and cheese. Dot liberally with butter, and pop into a medium oven until the casserole is heated through and the cheese is browned and bubbling. This dish can also be made quickly with cold mashed roots cooked the day before.

Or try combining the mashed roots with bread crumbs, butter, salt and pepper, and a dash of chili powder. Work the mixture thoroughly, and form balls with it, about the size and shape of small meatballs. Let the balls sit in the refrigerator for a couple of hours, and bring them out ½ hour before cooking. Then roll in flour and fry quickly in very hot oil in a deep fryer. Drain and serve as a substitute for French-fried potatoes, with barbecued chops or steaks. Because the balls somewhat resemble potato puffs, they generally make a great hit with kids.

SILVERWEED
Potentilla anserina

The turnip-like flavor of silverweed roots has been sampled by our ancestors in Europe and North America for many generations. Often consumed in enormous quantity because of the lack of other foodstuffs, the plant is said to have filled many gaps when commodities such as potatoes and bread flour were scarce. But silverweed is not such a bad substitute. Even today, its somewhat sweet-potato-like flavor can be enjoyed alone or in combination with most anything on the dinner table.

Silverweed is an attractive plant. It spreads by sending out numerous runners along the ground which root at varying intervals, firmly se-

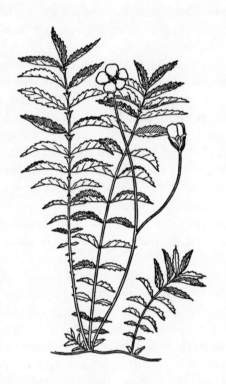

curing the plant to the soil. From the roots arise compound leaves, bearing numerous feather-like leaflets. The stolons and leaf stalks are covered with tiny, upright hairs. The oblong leaflets are a glossy green on their upper surface and a soft silver on the underside. One variety of the silverweed bears leaves that are silvery on both sides. The silvery leaves have been responsible for the French common name *argentine*. The plant is also known as wild sweet potato, wild tansy, and by a half dozen other names.

The flower of the silverweed is characteristic of all rose family members. The single, five-petalled blooms arise on long stems from the runners. About an inch in width, they are a bright, pretty yellow, and produce fruit that resembles dry strawberries.

Silverweed is a lover of damp places. It grows along stream banks and other gravelly shores, along sandy shorelines, and near stagnant ponds. It is mainly a northern plant, occurring from Newfoundland to Alaska and south to New England, the midwestern states, and California.

The thick, fleshy roots of the wild sweet potato can be eaten either raw or cooked. They are especially good candied, taking on a bit of the magical flavor of candied yams, but with a zing of their own. To candy young, firm roots, boil them in salted water until tender, and drain. Cut into any size pieces that you want. We leave the roots whole. Layer the roots in a buttered casserole dish with a ¼ cup of butter, ⅓ cup of brown sugar, 1 tablespoon molasses, 1 tablespoon of lemon juice, and ¼ cup water. Then bake in a moderate oven for ½ hour until the thick candy coating has adhered to all of the roots. For a delightful change, the brown sugar and molasses can be substituted with ½ cup of maple syrup and lemon juice to taste.

To make silverweed fritters, take 3 cups of cooked, mashed roots, and mix thoroughly with 2 tablespoons of butter, 2 slightly beaten eggs, salt and pepper to taste, and ½ cup of slivered, blanched almonds. If the mixture is dry (and it usually is not), add enough warm milk to make it moist enough to form patties. After making the patties, layer them in a dish, separating the layers with waxed paper, and let them age for an hour or more in the refrigerator. Then take the patties out, coat them lightly with flour, dip in beaten eggs, and coat with corn flakes crumbs. These patties can either be sautéed in butter in a non-stick frying pan until warm, or deep fried and then drained on paper towels. Either way, they go well with venison chops, a fresh salad, and a robust red wine.

POKEWEED

Phytolacca americana

Although several Indian tribes reputedly drank tea made from the berries of the American pokeweed as a cure for rheumatism, the practice is not highly recommended. Pokeweed must be collected with care for the table, because only the young shoots, under six inches in height, are edible. The roots and the mature purple stems are both poisonous, and the berries are known to cause reactions of sensitivity in some people, as well as some birds and wild animals. Some botanists believe the seeds to be poisonous. However, the young shoots make safe and delicious eating.

Young pokeweed shoots can be gathered in the spring of the year, from April at southern latitudes to June in the northern portion of the plant's range. Or, if you want a supply of fresh shoots throughout the fall and most of the winter, the plant can be cultured in a not-too-cold basement with a minimum of care. Medium-sized roots, pruned to four or five inches in length and planted in a deep flat of soil in a dark, cool place, will produce shoots for several months, and a crop can be harvested every ten days or so. About a dozen and a half roots can keep a

Young edible shoots

family in fresh shoots for a long time after the garden greens have given up for the year.

The genus name *Phytolacca* is thought to have been derived from the Greek word *phyton* meaning plant, and from the Latin *lacca,* referring to crimson-lake. The name aludes to the dark crimson stain that is left on anything coming in contact with the squashed or broken berries. The American pokeweed also bears the common names of poke, pigeonberry, scoke, garget, inkberry, and others.

A stout perennial of warm-temperate and tropical regions, pokeweed abounds from Florida and Texas, across New Mexico and Arizona, into southern California. Its range extends northward to southern Maine, southern Quebec and Ontario and the Great Lakes region. The plant favors rich, low, open ground, and can be found growing in newly established clearings, along roadsides, and on the borders of cultivated fields.

Pokeweed is easily recognizable. Its stout, coarse stems grow up to eight feet in height and become a deep purple with maturity. Large, alternate leaves arise on long petioles from the stem. The larger leaves grow up to ten inches in length, with wavy leaf margins and pointed tips. Opposite the leaves arise long, slender clusters of greenish-white flowers. Upon maturity, these give rise to clusters of blackish-purple berries, each one scalloped and flattened. It is useful to know the mature pokeweed plant, even though it is not edible, because the dried stalks remain standing in the spring following their fruiting, and the young, edible shoots always come up close by. So find one, and you'll always find the other.

To prepare young pokeweed shoots for the pot, remove the outer leaves and skin, and wash the whole stems thoroughly. The leaves can be cooked in a little salted water, drained, and smothered with butter. And for that matter, the stems can be treated in the same way.

The stems are also delicious boiled until tender like asparagus, drained, and placed in a buttered baking dish. Then dot with butter, sprinkle with buttered bread crumbs or crumbled corn flakes, mixed with sharp, grated cheddar cheese. Bake in a moderate oven until the cheese has melted and the crumbs are browned.

Pokeweed stems are good when boiled, drained, and covered with a thick, creamy sauce. For the sauce, melt 3 tablespoons of butter and stir in 4 tablespoons of flour and a dash of salt and pepper. Then add 1½ cups of game bird stock (or chicken stock if the other is not available), and slowly bring to a boil. Boil gently for a couple of minutes, then add ½ cup of cream and 2 egg yolks, well beaten. Add a touch of mace and lemon juice, and pour over the hot stems. If you like, decorate with slivers of blanched almonds.

PASTURE BRAKE
Pteridium aquilinum

After a long winter of shoveling snow, driving in blizzards, and fighting the traffic on the downhill ski tows, spring is a welcome affair. For the outdoorsman, a new generation is born; a new year begins. For the wild harvester, a great abundance and variety of wild edibles appears, free for the taking. And after a long winter of eating frozen vegetables and carrots from the root cellar, what better heralder of spring is there than the young fronds—the fiddleheads—of the fern family?

The pasture brake, known alternately as bracken, brake, bracken fern, and by a half score of other names, is the most common and probably the best known of the ferns. The young, tightly-coiled fronds of this plant make a delicious vegetable dish. But the young fiddleheads, named because the uncoiled stalks and leaves resemble the tuning end of a violin, are the only parts of the plant that are edible. The older fronds, after the tips have begun to straighten, become very tough, and mature fronds have been known to be detrimental to the digestive tract.

Some botanists believe the old fronds to be poisonous. This seems to be the case, though only when the old stalks are eaten in great quantity— like a plateful every day! The young fiddleheads are completely safe in any amount.

Pasture brake is a plant of dry, open, and oftentimes sterile woods. It is a frequent inhabitant of burns and other clearings, and occasionally appears in pastures and small thickets. The plant ranges from southern Alaska all the way to Mexico, and eastward across the Rocky Mountains and the midwestern states to Newfoundland.

The pasture brake is a coarse, stout fern. It has long been favored by the Japanese as a soup ingredient, and as a result the plant has been extirpated in many parts of that country. As a measure against extinction of the species, stringent laws have been passed to afford the fern the protection it needs. In North America, the plant is so far very abundant.

Early in the spring, the dark, perennial roots of the pasture brake, give rise to young fronds which are gathered by pinching off with the fingers as far down as possible, avoiding the tough bottom tissue. Later in the spring, the fronds unfurl to form erect, tall stalks with delicate leaflets typical of the ferns. The uncurling fronds are tri-forked, and usually sport a dark spot at the angle of the forks. The old stocks dry and usually become trampled down by the snows of winter, but are often found in spring close to the young fiddleheads.

Pasture brake fiddleheads can be eaten either raw or cooked. They add a bit of zip to a salad, but because some members of our family have never been overly enthusiastic about the mucilaginous quality of such vegetables as okra (bracken fiddleheads have the same quality), the young fronds are cooked and cooled before hitting the salad bowl. To prepare fiddleheads for the table, rub them briskly in your hands or draw your hand over the length of them to remove the loose brownish fuzz that covers them. Then wash thoroughly in water.

The simplest (and one of the best) recipes for fiddleheads involves simmering the heads until tender in a small amount of water. This does not take long. Then salt and pepper to taste, and smother with butter. Any leftovers can be sprinkled with a little lemon juice or wine vinegar and salad oil and eaten for lunch the next day. Even better, heat the leftovers slowly in a non-stick skillet in butter with a small amount of chopped, mild onion. This makes a great breakfast companion for eggs and toast.

For a real treat, try fiddleheads on toast, covered with a rich lobster sauce. For the sauce, cover lobster shells with a little water and cook in a pressure cooker for about 15 minutes. Strain the liquid, and season it with salt, pepper, and lemon juice. Add 1 cup of this juice to 2 tablespoons of butter and 3 tablespoons of flour that have been heated and blended together. Then bring to a boil and simmer for 5 minutes. Add ⅓ cup cereal cream, 1 beaten egg yolk, and pour over the hot fiddleheads on toast.

Pasture brake can be an important emergency food in the spring months. The rhizome can be eaten if necessary, and indeed was an important foodstuff of many Indian tribes. The dried, pulverized rhizomes provided flour for bread, and even the white core of the rhizomes was baked and eaten. Like the Japanese, the Indians used the young fronds to thicken soups and stews as well.

OSTRICH FERN
Pteretis nodulosa

Although its distribution is not as wide as that of the pasture brake, nor is it as well known, the ostrich fern is the plant referred to by many people when they say "fiddleheads." Indeed, in our part of the country, the term fiddleheads refers solely to the young, tightly coiled fronds of the ostrich fern, and is not generally used for other ferns or brackens.

The ostrich fern, *Pteretis nodulosa,* is alternately classified in various botanical texts as *Pteretis pensylvanica, Onoclea Struthiopteris,* and *Matteuccia Struthiopteris.* The last specific name is perhaps the most commonly used.

In contrast to the mucilaginous texture of the pasture brake, ostrich fern fiddleheads are quite dry. They can be collected in the same manner, by pinching the young fronds off as low as possible, or by cutting with a small knife, without getting into the tough tissue. The young croziers (this name is thought to be derived from the resemblance of

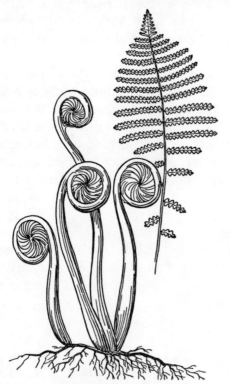

the young fiddleheads to the staffs of bishops) must be washed thoroughly, a n d their dry, paper-like scales removed. Once the fronds have started to unfurl, they are no longer edible.

The ostrich fern bears a multi-branched, perennial rootstock, spreading vigorously underground. F r o m the rootstock arise dense, vase-like clumps of young fronds, e a c h with thick stems. Last year's old, dried fruiting fronds are usually not too far away. The new fronds bear broad scales, papery and brown in color, and have a delicately leafy top.

Inhabitants of very rich soil areas, ostrich fern fiddleheads can be sought in stream beds, on fertile slopes, and in rich-soiled valleys and fields. The plant occurs across the entirety of southern Canada, from Atlantic to Pacific, and south to Virginia, the Great Lakes states, and the American Midwest.

Simply boiled, seasoned with salt and pepper, and smothered in butter, ostrich fern fiddleheads are delicious. They are even better boiled in a little salted water, drained, and placed in a shallow, oven-proof serving dish. Then dot with butter and sprinkle with a fifty-fifty combination of grated cheddar and Parmesan cheeses. Pop under the broiling element until the cheese has melted, and serve right away.

For an interesting casserole, mix 3 cups of finely chopped (preferably minced) cooked ham, 2 beaten eggs, 3 tablespoons mayonnaise, 1 teaspoon prepared mustard, salt and pepper, and 1 cup of grated cheddar cheese in a large bowl. The mixture should be moist and stick together quite well. If not, add a bit of milk or cream. Butter a casserole dish and line it with boiled and drained lasagne pasta. On top of the lasagne pasta, place a layer of the ham mixture. Then cover with a layer of boiled and drained fiddleheads. Over the ferns, pour a thin layer of white sauce—just enough to keep the contents of the casserole from drying during baking. Then start layering in the same manner with pasta, ham mixture, fiddleheads and white sauce, finishing with the sauce. Bake in a moderate oven until the casserole is heated through and set. Then quickly sprinkle with a little grated cheese and place under the broiler to brown. Spruced up with sprigs of parsley and thin slivers of sweet red pepper, this casserole makes a mighty fine Saturday lunch.

GLASSWORT

Salicornia spp.

Glasswort is a plant of the salt marshes. Because of the environment in which it grows, it has a high salt content and tastes distinctly salty when eaten. Thus the plant should be given special "no-salt" treatment when being prepared for the table.

The glasswort has many common names, mostly descriptive of its growth habit and its uses. The name "glasswort" is indicative of the high soda content of the plant, and on its one-time use in the manufacture of glass and soap.

Glasswort is a very fleshy herb. Its stems are highly branched and conspicuously jointed. Although the plant is leafless, it does have small, almost unnoticeable scales that adhere closely to the stem. Under the scales hide the tiny, easily overlooked flowers. The leafless, branched stems of glasswort sometimes resemble the foot of a barnyard chicken, hence the alternate names chicken-claw and pigeonfoot.

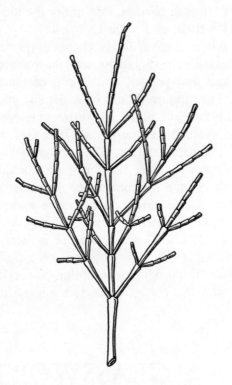

In summer the stems of glasswort are a bright, shiny green. In autumn, they turn to various shades of yellow, orange and red. Along with the change of color appear tiny, numerous seeds much sought after by many species of waterfowl. Indian tribes of Utah and Nevada also sought the seeds to grind them for flour.

Several species of glasswort thrive in the coastal, saline regions of this continent. They grow mostly in salt marshes, sandy seastrands, and in low, alkaline soils. Some also occur locally in inland salt licks and brackish marshes. The range of the plant covers both Atlantic and Pacific coasts from Alaska and Labrador right down to the Gulf of Mexico. Because of the salty environment in which the glasswort lives, and also its taste, the plant is sometimes known as saltwort.

Yet a fifth name for the glasswort, derived from the fact that it is best used as a pickle, is samphire. It is not, however, the same plant as the samphire in Europe. Pickles are most commonly made by first boiling glasswort stems until tender in a small amount of unsalted water. It is important not to use any salt, because the plant supplies its own. Once the stems are tender enough to pierce with a fork, they can be drained and packed in glass canning jars. At the bottom of the jar, if desired, you can place one onion sliced thinly and a clove of garlic. Then, over the top, pour mild vinegar that has been heated to the boiling point with a bag of pickling spices. Seal and store for a couple of months — no sampling beforehand!

Dilled glasswort pickles are also a favorite, especially with nibbles of cheese, crackers, and smoked whitefish. To make these, try to select stems that are juicy and young. These need not be boiled before pickling. Soak the stems overnight in very cold, unsalted water. When morning comes, drain them and pack them, along with a couple of sprigs of dill and a clove of fresh garlic, in canning jars. Then pour hot vinegar, mixed half-and-half with water, over the greens, and seal. In a couple of months everybody is eagerly waiting for the jars to be opened.

Although mainly used as a pickle, glasswort stems also impart a certain salty flavor to mixed salads that makes a salad memorable. However, you must remember to make a dressing that is salt-free, otherwise you will be at the beer barrel all night long.

Saltwort is an important species to remember for your list of edible emergency plants, because it is easy to identify and prolific in its haunts. It can be eaten raw, but its salty taste will send you running for fresh water. In the wilds, it is best boiled in a couple of changes of unsalted water.

Chapter III

NATURE'S STARCHES

Potatoes and rice have long been staples in our pantries. They can be prepared for the table in any number of ways, and provide both bulk and nutrition when served with the hot meal of the day. However, they can become a bit monotonous with time. And when this happens, most cooks have to delve deep into the recipe files to try to pacify the hungry hordes that appear each night at dinner time. This is where the wild edible starches come into their own.

There are numerous starchy plants that grow on this continent, in the pastures, marshes, and ponds right near home. Most are relatively easy to harvest, and require little or no special preparation to ready them for the table. Some have distinct flavors and require some getting used to, but the majority can be enjoyed right from the first taste. In addition, many of the wild starches are very low in fats and high in vitamin content.

The cooking possibilities for nature's potato substitutes are almost endless. Although a number of recipes are given in this chapter, the cook is at liberty to try anything that sounds appealing. For instance, how about French fried Jerusalem artichokes with a thick, juicy, barbecued steak? Or scalloped arrowhead tubers with a tender veal roast? Try stuffing your next chicken or duck with a wild rice dressing. You will find that cooking with the wild starches opens new gates, and presents many new taste experiences.

CATTAIL

Typha

Almost every child that has had any exposure to the outdooors, and particularly to marshy or swampy areas, knows the distinctive and haughty cattail. Four species of cattails, or bulrushes as they are oftentimes called, proliferate throughout the wet areas of North America. All are members of the genus *Typha,* and all make good eating.

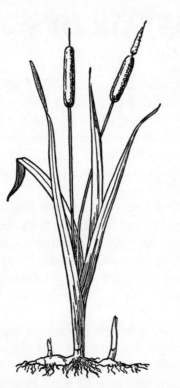

Typha latifolia, the broad-leafed or common cattail, is probably the best known of the four. It lives in marshes and shallow water areas from Newfoundland to Alaska, and south through much of the United States into Mexico. The narrow-leafed bulrush tends to like warmer weather, and shies away from the cold areas of central and northern Canada. The remaining two species are even more sun-loving and are quite at home in the sunny south and the tropical Americas.

Cattails are most readily recognized by the novice plant hunter beginning in July and August, when the long and dense cylindrical spikes of flowers are formed. The flowering stocks — the rushes or flags — hardly look too appetizing once September comes around and they become very dark in color; but tribes of Paiute Indians are known to have eaten them early in the summer, before any pollen became evident. We have tried it on a number of occasions and have found good eating in young flower spikes, quickly boiled like young corn cobs, and liberally smeared with butter and a little salt.

Once the flower ages and pollen appears, the flowers can be harvested for their golden yellow pollen. A plastic bag will work wonders here. You can remove a flowering stock (with a hefty knife or a pair of secateurs or pruning shears) and shake the flowers vigorously in the bag. If you are tall enough, just shake and leave the plant intact. The golden yellow, dusty particles that result are excellent when combined about one to two with flour and used for pancakes, breads, or puddings. The pollen gives off a marvelous yellow hue in baking bread that makes you think a half dozen egg yolks must have been used.

The flowering spikes of the cattail are not the only edible part. Young rootstocks (actually rhizomes or underground stems) are quite tasty when prepared with the outer skin removed. The interior is very white and quite firm, like most underground plant parts. Rhizomes form a tangled mass under the surface of the marsh, giving rise to roots at intervals along the way. Indians dried and ground the rhizomes, using them as flour.

Back in spring, when young cattail shoots are just beginning to emerge, the shoots can be eaten like any young and tender vegetable. They are good simply boiled in water and covered with butter, a little salt, and pepper. But they are really superb when smothered with a rich, thick, cheese sauce. We remove the outer leaves, boil the young shoots in lightly salted water and cover them with a basic white sauce to which we have added ½ pound of grated medium cheddar cheese, ½ teaspoon of dry mustard, and a few sprigs of fresh parsley, finely cut.

The Blackfoot Indians used a pack of cattail down as a burn treatment; other Indian tribes used the down in bedding, pillows and comforters. More recently, floral designers have found the line and texture of young flowering stems of *Typha* useful for dried flower arrangements. Cattails have always been excellent survival rations, because they are easily recognized and can be eaten both raw and cooked throughout most of the year. But not too many people have pulled on their rubber boots and headed out into the marsh in recent years to harvest wild cattails for the table. Go out this year and try it. It tastes like asparagus, but is much cheaper than the supermarket stuff!

WILD RICE
Zizania aquatica

Few plants have the history or the heritage of wild rice. In the marketplaces of eastern cities, the grain commands a price that is somewhat unbelievable. Yet its texture and flavor make it well worth seeking in areas where it grows in abundance and is free for the taking.

Indian tribes have long used wild rice as a cereal food. The Menominee tribe even takes their name from the plant. The Indian women's method of collecting the grains was an efficient one. On the day of ripening, the women went out in a canoe to the tall, plume-topped plants that they had tied together and bent some time before. They then beat the seed-bearing plants, allowing the grains to drop into the

bottom of the canoe. Grains that fell away from the canoe into the water ensured a future crop. The rice was then dehusked by parching over hot coals, stirring constantly to prevent burning. The cooled husks were later beaten off and the seeds winnowed.

A similar method of preparation can be used by today's wild harvester, but it is a bit simpler to just let the grains dry out in a warm, dry place for a few days. After the grains are dry, they can be parched on a cookie sheet in a moderate oven for a few hours. They should, however, be mixed and turned during the parching process, and watched carefully to keep them from burning. The husks can then be beaten or blown away.

Wild rice is very easy to recognize. It is a broad-leafed grassy plant, thick-stemmed, and reaching a height of up to ten feet. The flower head is plume-like, with the pollen-bearing flowers at the base of the plume and the seed-bearing flowers at the top. The seeds are slender, dark, about a half inch in length, and covered with a loose husk that sports a bristle-like hair at the tip. In areas where it is abundant, it is almost impossible to mistake for anything else.

Wild rice grains ripen for harvest in midsummer and early fall. But you have to keep a close eye on their ripening progress, because they drop very quickly once they are ready.

Water oats, as wild rice is sometimes dubbed, is an annual plant, proliferating in quiet waters, fresh and brackish water marshes, and in tributary streams and river mouths. There are several varieties of *Zizania aquatica,* differing mainly in the width of the leaf and the height of the plant. Wild rice ranges from southern Maine and Quebec to Wisconsin, Minnesota, and the Great Lakes area, and south along the Atlantic seaboard to northern Florida. The inland variety of wild rice grows from Indiana to North Dakota, and south to Missouri and Texas.

There are almost as many recipes for wild rice as there are grains harvested by the Indians in northwestern Ontario. Every chef has his or her own specialty. Wild rice can be used in any recipe calling for long-grain white rice, but using it indiscriminately can become expensive if you are buying it and time-consuming if you are doing the collecting yourself. Our favorite uses for wild rice are with game.

Young wild mallards stuffed with wild rice dressing are a delight. We take about ¾ of a cup of wild rice grains, wash them thoroughly in cold water to remove some of the heavy, smoky taste, and soak them in cold water overnight. Then we boil the grains until tender. This makes about 3 cups of boiled rice. We then sauté half of a finely chopped sweet pepper, a small finely diced Spanish onion, a stalk of chopped celery, a few sprigs of chopped parsley, and salt and pepper to taste, in butter. To this we add the rice and 1 cup of bread crumbs, mix it all together, and stuff our mallards. With a salad of fresh greens and a robust red wine, it makes a marvelous dinner.

To stretch our wild rice supply throughout the winter and spring, we mix it half and half with long-grain white rice. The smoky wild rice flavor still comes through nicely. Sweet green peppers stuffed with wild rice and mushrooms are another of our favorites.

Wild rice is very low in fat, high in protein, and rich in vitamin B. The Indians recognized its high nutritive value, and thanked the Great Manitou in yearly thanksgiving ceremonies for his gift of the grain. We, too, should be thankful for this delicious wild edible.

JERUSALEM ARTICHOKE

Helianthus tuberosus

The Jerusalem artichoke, a member of the large sunflower family, is somewhat misnamed, as it is not a true artichoke and does not grow in Jerusalem. The plant is indigenous to the central portion of North America. It was widely cultivated by the Indians, and introduced into Europe early in the seventeenth century. The tubers were called *girasole* in Italian and *girasol* in Spanish, meaning sunflower, and these names are said to have been misinterpreted by the English to produce "Jerusalem." Hence the name stands today.

The Indians ate Jerusalem artichokes raw, boiled, and baked. The

plant quickly escaped from cultivation, and s p r e a d across almost the entirety of the continent. The plant proliferates in rich and damp thickets from Ontario to Saskatchewan, and south to Georgia, Tennessee, and Arkansas. Moist soil a r e a s along stream banks, ditches, roadsides, and fencerows are good places to look for this sunflower.

The Jerusalem artichoke is a rough-looking plant. It is perennial in nature, and reaches a height of up to ten feet. The stems are coarse and hairy; the leaves, thick and hard, with rough hairs on the upper surface. The numerous flowers a r e a light, pretty yellow, without the prominent brown centers of the other edible sunflowers. The flowers are two to three inches in diameter, and are not particularly striking. The saving grace of this large, coarse sunflower is its tubers. These enlarged, underground stem portions have a slightly sweet and most distinctive flavor. They are somewhat watery in texture, and make a good wild substitute for potatoes.

The Jerusalem artichoke is an excellent food and cover crop for pheasant and quail flypens. European game keepers have used it for decades, and it is becoming more popular with North American game-bird breeders as well.

Some people don't care for the taste of the Jerusalem artichoke. But it is something that grows on you with exposure. We have tried a variety of different recipes, and although my husband has come to like the vegetable, my father-in-law still says that he likes his best second-hand — after the pheasant has eaten it!

Jerusalem artichokes are quite nutritious and can be enjoyed by invalids because of their easy digestibility. The fleshy tubers are reported to be free of starch, so that gives them a decided advantage over potatoes as table fare.

The tubers should be collected in late fall. They can be simmered in water with their skins on and then cooled and the skins removed, or they can be peeled before boiling. Boiled and smothered with butter and a little salt and pepper, they are quite tasty. As you are peeling the tubers, soak the finished ones in water with a little vinegar or lemon juice to keep them from turning brown. Scalloped artichokes, following a recipe for scalloped potatoes, are also very good.

We like fried artichokes, and cut and deep-fry them just like French fried potatoes. We also use them as a fondue ingredient, dipping them in hot oil for about 10 minutes along with mushrooms, Spanish onion slices, and chunks of beef or moose. For a fondue ingredient, we peel the tubers, cut them in small pieces, and put them on the table in mildly acidulated water.

Jerusalem artichokes are also good in salads. Our favorite artichoke salad uses thin slices of the tubers, tossed with lettuce, tomatoes, onions, green peppers, and an oil and vinegar dressing liberally laced with crushed, dried marjoram, chopped parsley, finely chopped tarragon, a bit of thyme, and a few drops of Worcestershire sauce.

ARROWHEAD

Sagittaria spp.

The tubers of the arrowhead were long-time favorites of a number of Indian tribes across North America. Women had the job of gathering the bulbs of this water weed, and did so while holding on to a canoe for support. Sometimes submerged neck-deep in water, they worked the tubers loose from the wet ground with their feet. The freed tubers floated to the surface, and were gathered and tossed into the bottom of the canoe.

Fortunately, most of the seven or so species of *Sagittaria* with tubers large enough to be harvested grow in shallow water areas where collecting them does not involve a canoe or neck-deep water. The arrowheads are plants of wet places such as ponds and fresh-water river margins. Their range extends from coast to coast, and from southern Canada to Louisiana, Oklahoma, and on into Mexico.

The *wappato*, as the arrowhead was called by the northwestern Indians, runs from a foot to three feet in height, and is characterized by rosettes of erect leaves on long petioles, appearing from the roots. The long, arrow-shaped leaves vary greatly in size and shape, depending on the variety of the plant. Most do resemble an arrow or a lance. The flowers are easiest to recognize. Circles of flowers appear from July to September, near the top of a naked flower stalk that stands a head above the leaves. The flowers are white and fragile in appearance, usually in threes, and are of two different types — gold-centered pollen-bearing flowers near the top of the spike, and green-centered seed-bearing flowers lower down. In fall, rounded heads of flat seeds are produced.

Arrowhead tubers are most easily collected by reaching under the muddy surface of the pond or river bottom and freeing the tubers or digging them with a small trowel. Excessive pulling at the tops of tall plants frequently results in yanking off the leaves and leaving the tubers safely in the ground.

Once collected, the hard potato-shaped tubers, which run about the size of an egg in some of the larger species, can be simply washed and eaten raw. They make an excellent emergency food for outdoorsmen. However, like potatoes, they are best cooked. They can be boiled (best done in a pressure cooker because they take longer than new, cultivated potatoes), jackets removed, and served with butter, salt and pepper. They can also be baked, scalloped, creamed, fried, roasted, or cooked in just about any other way you can imagine. One of our favorites is to boil the arrowhead tubers, cool and remove the jackets, then slice and slowly brown in butter with a dash of salt and pepper. Once the "potatoes" are thoroughly browned, we add ¼ of a cup of cereal cream and simmer to heat the cream. Served hot with venison chops, they have a delightful flavor.

Arrowhead tuber salad is also good, and can be made from any potato salad recipe. My husband likes hot arrowhead salad; I prefer mine cold with sour cream. We precook our tubers, remove jackets and cool. Do a whole batch at a time. It will disappear fast! When cool, we dice the tubers and add 1 to 1½ cups finely chopped celery, a few sprigs of chopped parsley, ½ of a finely chopped onion, a few chopped chives, salt and pepper to taste, 1½ cups of creamed cottage cheese, and a cup of sour cream. Now we let this sit in the refrigerator for a while (2 to 3 hours) while we tend to the Sunday afternoon football game, and serve it cold with dinner.

GROUNDNUT
Apios americana

In our household, one of the starchy vegetables such as potatoes or rice is expected on the dinner table with the main meal each day. And, of

course, it can't be the same old dish every evening. Variety is also expected, and good taste as well. One of the plants that provides all of these prerequisites, as well as high nutritional value, is the groundnut or Indian potato.

Not many plants enjoyed such tremendous popularity with both Indians and settlers alike as the groundnut. For generations, it was an important foodstuff nutritionally for many Indian tribes. Upon the arrival of the Pilgrims at Plymouth Rock, the red-skinned natives shared their knowledge of the plant with the newcomers, helping the settlers through their first, hard winter in the new world. Alas, as reports have it, less than twenty years later laws were passed forbidding the Indians to harvest groundnut tubers on English lands, on penalty of jailing, and whipping for a second offence. News of the tuber which the Indians called *hopniss* spread back to Europe where several attempts at cultivation took place, but the idea was soon abandoned as impractical.

The perennial American groundnut is a thin, twining and trailing vine. Its slender, soft stems give rise to alternate leaves, each bearing five to nine narrow, oval-shaped leaflets with short stalks. In winter, the stems take on a whitish hue as they dry. In late summer, the leaf axils sport thick clusters of brownish to purplish flowers. As a member of the pulse or pea family, the flower clusters strongly resemble those of domestic peas and beans. The blooms have a rich, heady sweetness of aroma that is hard to forget. Later in the year, pods are formed, again resembling bean or pea pods. These are borne more or less in clusters, and the "peas" can be harvested and eaten much like the cultivated varieties. The Indians are known to have harvested the pods to eat the peas.

The most significant part of the groundnut, however, for the wild harvester is the root. The rootstocks of this plant resemble a string of beads, with numerous tuberous enlargements strung together in a chain. These tubers or "nuts" vary from one to three inches in length, and grow not too far below the soil's surface. The tubers, as well as the stems, bear a milky juice.

The groundnut is an inhabitant of low ground and rich soil areas. It can be found in rich thickets and near streams and brooks. An eastern species, its range covers most of the eastern United States and southeastern Canada, from New Brunswick and southern Nova Scotia through New England to Florida. Westward, its range stretches to Texas, Minnesota, and Colorado. The groundnut is also reputedly native to China.

A cousin of the American groundnut also grows in woods and thickets in Tennessee and Kentucky. *Apios priceana* bears a strong resemblance to the common groundnut, but develops only a single, irregularly shaped tuber that runs three to six inches in diameter.

Although raw groundnut tubers are edible, they are not what one would call choice. They are a little on the tough side and the milky sap is a bit disagreeable to the palate. The tubers are delicious when cooked and piping hot, but again lose their appeal when cold. Despite the name of Indian potato, the flavor of the "nuts" more closely resembles that of rutabagas.

The simplest, and one of the tastiest, methods of preparing the tubers for the table is to boil them slowly, unpeeled, in salted water until they are tender enough to be pierced easily with a fork. Then drain and smother with butter. Sour cream with chopped chives can be served in a side dish as a topping for the buttered tubers. They are also good when parboiled and then tossed into the bottom of the roast pan to finish cooking with the roast, onions, celery, and carrot sticks. Any leftovers can be sliced and quickly pan fried in butter for serving.

Groundnut tubers are thoroughly enjoyable fried. To prepare the tubers for frying, wash them carefully and slice them, skins still on, about ⅛-inch thick to give thick potato-chip slices. Melt 6 tablespoons butter in a non-stick frying pan. Sauté the tuber slices until they are tender and thoroughly cooked. They should take on a light golden color. Remove to a serving dish, sprinkle with salt and pepper, and serve immediately.

SPRING BEAUTY

Claytonia spp.

Where the plant grows in abundance, spring beauty can provide another delightful change from potatoes and carrots. However, despite its liberal range which extends over the majority of North America, this member of the purslane family does not always grow in profusion. The deliciously edible roots should be harvested with preservation of the species in mind. This precious and delicate wild flower, as its name indicates, adds unlimited beauty to the spring woods.

Four species of spring beauty, or fairy spuds as they are also commonly known, are believed to grow in North America. The Carolina species, *Claytonia caroliniana,* is an inhabitant of rich, alluvial soils, making its home in open woods, slopes, and thickets. It is most common from Newfoundland to Saskatchewan, through southern Nova Scotia and New England, and into the middle eastern and midwestern states. The Virginia species, *Claytonia virginica,* inhabits similar terrain, over a somewhat larger portion of the continent, including the south. Several sub-species of the Virginia form exist. The spring beauties are also found locally in the mountainous regions of the West and as far north as Alaska.

A close relative of miner's lettuce, spring beauty leaves are high in vitamins A and C. The leaves are somewhat thin and grass-like, sometimes broadening to succulent, ovate leaves depending on the species. The leaves arise opposite one another on the stem. The plants grow up to about ten inches in height.

In the spring of the year, masses of white to pinkish flowers provide a spectacular sight where the plants are plentiful. The small flowers grow in clusters, but arise from a small tuber. The stems of the plants are often too weak to support a lush growth of blooms, and often lean over under their weight.

The small, starchy roots are the part of the plant that is best enjoyed on the dinner plate, although the young leaves can be eaten either raw or cooked. Reputedly favored by a number of Indian tribes, the roots can be dug with a small trowel or any other self-fashioned digging stick. They must be thoroughly scrubbed and washed before cooking.

The roots are enjoyable simply boiled in salted water, unpeeled,

until tender enough to pierce with a fork. Then drain, remove the jackets, and drench with melted butter with some chopped parsley added.

Whenever we are fortunate enough to be able to harvest a small quantity of spring beauty roots, we like to savor them in a moose stew. Any stew recipe will do. Ours contains tender moose stewing meat, mushroom, onions, cubed carrots, diced sweet peppers, chopped celery, and whole spring beauty roots. Salt and pepper are the only seasonings, as the stew ingredients blend their flavors nicely to give a rich, tasty dinner. We purposely make the stew a little watery so that we can cook fluffy dumplings, made with sesame and celery seeds, in the stew juices just before serving.

WATER LILY
Nuphar spp.

Studying the taxonomy of the water and pond lilies is a somewhat confusing experience. There are many of them in North America — some eastern, some western, others occurring over most of the continent. The common names change rapidly with locality. But basically the water lilies are all perennial, aquatic species with spongy rootstocks and urn-shaped fruiting heads. Most children that live near a pond or quiet stream can tell you what the plant looks like. The big, globular flowers

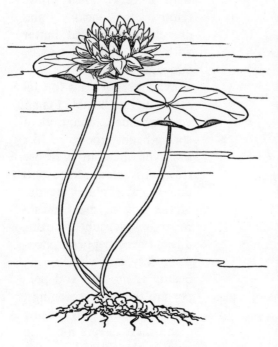

are favorites with everyone. The northwestern species of water lily, *Nuphar polysepalum,* has a rich history of Indian usage. It is said that ceremonial dances were held by the Klamath Indians at the time of the water lily seed harvest. Once collected, the large pods were sun-dried and then pounded to loosen and remove the seeds. The seeds were then parched, pounded, and winnowed to remove the shells. The inner kernels were eaten roasted or w e r e ground into flour and stored. The rootstocks were also harvested. Boiled or baked, the inner core, which is very rich in starches, was enjoyed for its sweetish taste. The starchy rootstocks were sometimes also ground into flour.

The yellow or bullhead water lily, *Nuphar variegatum,* and the tuberous water lily, *Nymphaea tuberosa,* among others, are also popular among seekers of wild edibles. Not many people today collect the seeds, but they are reputed to make bread that is hard to beat, and kids delight in making campfire popcorn from the ripe seeds.

Aside from differences in flower color and a few other variations, most of the water lilies are very similar in character. Their leaves, which are thick and somewhat leathery, arise from the top of the heavy, spongy rootstock. They are borne on thick, also spongy leaf stalks, and are heart-shaped with deep basal notches. The leaves can grow to almost a foot in length in some species.

The leathery flowers are mostly large, and almost globular in shape. They range from yellow to white to various shades of off-white, oftentimes tinged with a greenish or purplish hue. The urn-shaped fruit houses numerous seeds.

The range of the water lilies extends over the temperate and cold-temperate portions of this continent from Labrador to the Pacific in Canada, north to Alaska, and south to California and the Gulf of Mexico. The plants prefer quiet waters, and are commonly found in pools, ponds, marshes, slow-moving streams, and bogs.

Although you may want to go to the trouble of trying out water lily seed bread, the rootstocks are much easier fare in the kitchen. The roots are quite easily gathered from the muddy water bottom by breaking them off with the feet or by rooting around with a canoe paddle. In the case of the tuberous water lily, simply walking around in the muddy water will break free enough tubers for dinner. The tubers readily float to the surface of the water once broken free of their matted, clinging roots. The tuberous roots can be peeled before cooking, or afterwards. When scrubbed and baked in the campfire in aluminum foil, they need not be peeled until eaten. If boiled, two changes of water should be used to alleviate some of the strongly sweet taste. Water lily roots are also good scalloped with milk and cheese, or tossed into stews.

EVENING PRIMROSE
Oenothera biennis

The evening primrose is a native North American. So relished was it by early explorers that it found its way very early in the history of this continent back to the Old World, where it was quite successfully cultivated. *Oenothera biennis* is the species most commonly recognized by wild harvesters, but the plant hybridizes so readily that many forms occur on this continent. These are considered by various botanists to be different varieties, races, and even different species.

The roots, shoots, young stems, and even the leaves are thought to have been used by various Indian tribes as foodstuffs. There is also some indication that the mild astringent qualities of the plant were looked upon as a cure for coughs due to colds.

As the name indicates, the evening primrose is a dusk bloomer. Its four-petalled yellow blooms remain closed from daybreak through afternoon, but open at sunset. The flowers are quite large and showy, and attractive after dark to the nocturnal moths upon which they de-

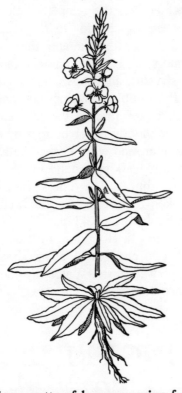

pend for fertilization. Come morning, the short-lived blossoms close their petals, begin to wilt, and eventually drop off. The flowers are borne in the leaf axils of the flowering stem.

As a result of the work of the moths, large capsule-type seed pods appear in the fall, each carrying many seeds. The irregularly shaped seeds are reddish-brown in color and find their way by means of oddly shaped wings to a resting place where they germinate and form another generation.

The evening primrose is a biennial with a large, fleshy taproot. The first year of growth produces a broad, flat-lying rosette of leaves varying from one or two inches to a half-foot in length. The leaves sport prominent midribs, pointed ends, and wavy or toothed leaf margins. In the second year, a flowering stem is produced. This stem stretches from two to six feet. It is hairy, sometimes branched, and becomes woody as the plant ages. The leaves on the flowering stem arise alternately, without leaf stalks.

The range of the evening primrose extends from Labrador and Newfoundland to British Columbia, and south to Florida, Texas, and the West Coast. The plant prefers gravelly, well-aired soils, and is common in dry fields, meadows, pastures, along roadsides, and in many waste places.

The stout roots, somewhat resembling those of a parsnip, are edible only when harvested in the first year of growth, before flower initiation and development take place. Actually, the roots could be eaten after that if circumstances necessitated it, but they would have to be boiled in at least three changes of water to get away from the toughness and the bitter, peppery taste.

The good old standby method of preparing evening primrose roots for the table is to peel the roots, cut them into bite-sized pieces, and then boil them in two changes of water. Sprinkled with salt and smothered with butter, they make a good starch substitute.

Once the roots have been peeled and parboiled, they can be enjoyed in any number of recipes. We like them in soups, both broth-type and creamed, as well as in stews. We also enjoy them finely chopped in omelets or scrambled eggs. Their lightly peppery taste adds a lot of zest greased pizza pans or one extra-large greased cookie sheet.

One young friend, who took an interest in wild edibles and spent a few Saturday mornings and evenings after school collecting wild species with us, came up with the younger generation's answer to the wild harvest—evening primrose pizza! It sounded feasible, so we tried it. Any recipe for pizza dough will do. Ours goes like this. In 1 cup of lukewarm water (about 110°F), dissolve ½ teaspoon of sugar. Add 1 level tablespoon of dry granulated yeast and let stand for 10 minutes. (Any active yeast can be used — just follow the accompanying instructions.) Add 2 tablespoons vegetable shortening, ¼ teaspoon salt, and 2 cups flour. Beat with a wooden spoon until thoroughly smooth. Gradually add 2 more cups of flour, mixing with your hands until the dough is sticky enough to form a ball. Knead the dough on a pastry sheet for 10 minutes. Then place the dough in a lightly greased pyrex bowl, cover with a towel, and let rise in a warm place for 1 hour. Punch the dough down and knead again for 5 minutes. Now stretch the dough to fit two greased pizza pans or one extra-large greased cookie sheet.

Cover the dough with any tomato sauce that you wish — either tinned or homemade. Over the top of the tomato sauce, place a layer of thin slices of mozzarella cheese. Then sprinkle with diced sweet green peppers, mushroom slices, thin slices of pepperoni, slivers of Spanish onion, and diced, precooked evening primrose root. Bake the pizza in a moderately hot oven for about a half hour, until the dough is cooked and the filling is bubbling. Magnifico!

Chapter IV

FRUITS OF THE WILDS

For the tyro wild harvester, the fruits of the fields and forests are often the most frequently sought wild edibles. And for several good reasons. The fruits are quite often the most conspicuous part of the plant, and hence the most easily recognized. They are also generally the most flavorful plant portion. And on top of this, they normally require little in the way of kitchen preparation before they can be enjoyed at the table.

Wild fruits are found everywhere in North America, from the tundra of the Arctic to the Gulf Coast, and from the tidal marshes of Newfoundland to the sandy beaches of California. At times they grow in great abundance, and require little effort to harvest. But perhaps their greatest appeal lies in the variety of ways in which they can be prepared as desserts, preserves, condiments, and snacks.

Some of the wild fruits are very well known. Wild blueberries, for instance, are easily recognized by those who seek them, and can readily be enjoyed either out-of-hand or in a variety of dishes, including the incomparable wild blueberry pie. Wild strawberries have long been enjoyed in jams and cakes. And applesauce made from wild apples has always been a favorite. But many fruits exist that are not as well known, yet are just as delicious and easy to harvest. Try pawpaw, for instance, or wintergreen, or mulberries. All are full of flavor and nutritious benefits, and all are free for the taking.

BLUEBERRY

Vaccinium spp.

What better way to begin a chapter on wild fruits than with wild blueberries — everyone's favorite. Wild blueberries make some of the best pies, syrups, and waffles that ever hit the dining room table, and they grow in such abundance in some areas of this continent that they are no trouble at all to harvest when the peak of the season is at hand.

The blueberries belong to the genus *Vaccinium,* a large category of plants encompassing all forms of blueberries, bilberries, and cranberries, a m o n g others. The blueberries differ from the bilberries in the location of the fruit on the plant. The true blueberries bear their fruit in terminal clusters, while the fruit of the bilberries, or whortleberries as they are sometimes called, arises from the leaf axils. Some twenty or more species of blueberries are recognized, all of which are edible, although some taste considerably b e t t e r than others, especially when eaten raw.

Two main types of blueberries grow in North America—those that grow on tall or high bushes and those that grow on short or low plants. High-bush blueberries, of which several species exist, grow on plants ranging up to twelve and fifteen feet in height. At times these big plants bear such an abundance of fruit that a whole basket of berries can easily be gleaned from one bush. The berries vary in color from frosted blue to purple and sometimes blue-black. The leaves of the common high-bush blueberry, *Vaccinium corymbosum,* are generally green on both sides, broad, and not fully developed when the whitish to pinkish flowers are in bloom. The fruit is juicy and sweet, and ripens from late June to September. The plant is an inhabitant of swampy areas, low, acidic soils, and occasionally dry woodlands. Mainly a southern species, the plant is common in the southeastern United States. However, its range stretches northward as far as Maine, Nova Scotia, southern Quebec, and Wisconsin.

The low-bush blueberries consist of quite a number of species, three of which are probably most commonly known among wild harvesters. One of the sweet low-bush types, *Vaccinium angustifolium*, has bright green, smooth leaves on both sides. The leaves are quite narrow, oblong in shape, and have a border of fine bristles. The fruit is blue and covered with a whitish or greyish bloom. This blueberry also favors acidic soil, and grows near peat bogs and rocky terrain. It is common in recently burnt-over areas. Its range extends from Labrador and Newfoundland to New England, and west locally to Minnesota.

Another sweet blueberry, *Vaccinium vacillans*, is characterized by a green uppersurface of the leaves and a duller or whitish undersurface. The leaves of this species are more oval in shape. The berries are a dark blue, with a faint bloom, and very sweet. A more southerly species than the first sweet low-bush variety, the plant inhabits dry open woods such as thin oak stands, clearings and thickets. It is found from Georgia and Missouri as far north as southern Nova Scotia and Maine, and Illinois to Iowa.

The third low-bush blueberry is not a sweet one. In fact, its sour taste has earned it the name of sour-top blueberry, *Vaccinium myrtilloides* (or *canadense*). It is easy to recognize because of its downy twigs and leaves. The small, smooth-edged leaves have also earned it the name of velvet-leaf blueberry. A moist woods and swamp dweller, this northerner grows locally from Newfoundland to Manitoba, through the northern states, and in mountainous country as far south as Virginia.

Many other species of blueberries and bilberries exist, but to describe them all even briefly would require a chapter in itself.

Members of the heath family, the blueberries are favorites of a number of wild creatures. Collecting the berries is easy, but cleaning them is another matter. All of the bits of leaves and stems must be removed. Yet however time-consuming it may be, wild blueberries on the table are well worth the effort.

We enjoy wild blueberries simply sprinkled with a little sugar (if they are not already sweet enough) and covered with cereal cream. We also like them for breakfast atop a non-sweetened dry cereal, again with light cream.

Stewed blueberries are a nice treat. To stew the berries, cook them slowly over low heat, with as little water as possible, until they are soft. Sweeten to taste, if necessary. Serve warm with cream, ice cream, or a heaping spoonful of sour cream.

Perhaps the best way to enjoy wild blueberries is in a pie. There are a number of recipes for the delicious wild berry filling, but usually the simplest ones are the best. For a fresh-fruit glazed blueberry pie, start with a baked pastry shell. Clean and dry enough berries to fill the shell, and sweeten if necessary. Then cover with a glaze made by mixing 1 cup of sugar, 3 tablespoons cornstarch, a dash of salt, and 1 cup of water with a teaspoon of lemon juice added in a saucepan. Heat this glaze mixture over low heat until it thickens. Then pour over the fresh blueberries in the shell. Chill the pie thoroughly and serve with freshly whipped cream.

For those who prefer old-fashioned blueberry pies with a cooked filling, two pie crusts are needed. Fill an unbaked pie shell with 3½ cups of cleaned blueberries, ¾ cup of sugar (adjusted to suit your taste), 2 tablespoons flour, and a dash of salt, all mixed together. Dot with butter, and sprinkle with a few drops of lemon juice if desired. Then cover with a top crust, flute the edges, and cut vents in the crust. Bake in a moderately hot oven (425°F) for 30 to 45 minutes, until the crust is golden and the filling bubbling hot. Incidentally, it helps when baking blueberry pies to either have an oven protector for catching spills, or a self-cleaning oven!

These are only the very basics of cooking with wild blueberries. They make almost any fruit recipe take on a special character all of its own.

HUCKLEBERRY
Gaylussacia spp.

The huckleberries are extremely close relatives of the blueberries, so close that oftentimes their common names are interchanged, they are picked together during harvesting, and they are mixed when being prepared for the table. However, anyone who knows both groups of fruits can tell you that the huckleberries are more seedy. Actually, blueberries contain many small fine seeds that don't make their presence conspicuous. They don't get caught in one's teeth! Huckleberries, on the other hand, have ten hard and somewhat stone-like seeds in each berry. Yet despite the seeds, the sweet and slightly spicy flavor of the huckleberries makes them first-class wild fruits.

The true huckleberries can also be distinguished from the blueberries in that their leaves are dotted with tiny "glands" which make them slightly waxy to the touch. The young shoots and the blossoms also sport these waxy secretions.

The black huckleberry, *Gaylussacia baccata,* is probably the best known on this continent. It is a highly branched shrub, the branches very stiff and grey-brown in color. From one to three feet in height, it sports alternate leaves, oval to oblong in shape. One-sided clusters of pinkish or pale reddish blooms give rise to shiny black fruits, occasionally but not often covered with a bloom. The small clusters of sweet berries ripen from late July to September, depending on locality.

Black Huckleberry

The black huckleberry is the most northerly of the various species mentioned. Its range stretches as far north as Newfoundland, southern Nova Scotia, and Saskatchewan. Southward, it extends through New England to Georgia and Louisiana. The plant favors openings, clearings, and pastures, dry or sometimes boggy.

The other huckleberries are southerners. The dwarf huckleberry, *Gaylussacia dumosa,* does occur as far north as New Jersey and southern Newfoundland, but it grows in peat bogs and wet thickets, and is not generally sought in these areas. In the more southerly states, from Florida to Mississippi, it inhabits drier sandy soils and pinelands, where its spicy, sweet, black berries are more readily accessible to the wild harvester.

The dangleberry, *Gaylussacia frondosa,* strongly resembles the black huckleberry, but its sweet, juicy berries are dark blue with a bloom. It is a resident of dry woods and thickets from southern New Hampshire and New York south to Florida and Louisiana.

Huckleberries can be used in most any recipe calling for blueberries, if the nut-like seeds don't bother you. We prefer them mixed together about half and half. A wild huckleberry and blueberry pie is a delightful taste experience. This combination of fruits is also good stewed and topped with cream.

For a delicious hot dessert, pour 3 cups of the stewed fruit combination into a shallow, lightly buttered casserole. Sprinkle with a little sugar, nutmeg or mace, and lemon juice to taste. Then cover with a batter made as follows. Mix 1½ cups pre-sifted flour, 1 tablespoon baking powder, ½ teaspoon salt, and ½ cup sugar. Separately, beat 1 egg with ½ cup cereal cream (or milk if you want to keep the calories reasonable), and ½ cup of melted butter. Stir the liquid into the flour mixture until smooth, but do not beat. Pour this batter over the stewed berries and bake in a hot oven (400°F) until the batter is crusty and golden brown. Spoon into cereal bowls or deep dessert dishes and serve warm with whipped cream or just with cream poured over the top.

CRANBERRY
Vaccinium spp.

Also very close relatives of the blueberries are the cranberries, everyone's favorite accompaniment for Thanksgiving turkey. The commercial cultivars of cranberries that appear in most cranberry sauces are big and plump, but they have nothing on the wild types for flavor or color.

Three species of wild cranberries inhabit the northern and eastern portions of this continent. All are basically low, thick, evergreen shrubs. The mountain-cranberry, or partridge-berry as it is popularly known in parts of the Canadian East (*Vaccinium Vitis-Idaea*), is a creeping evergreen with slender stems and branches. Its narrow, leathery, box-like leaves are smooth and bright above, but have duller, slightly mottled undersides with stiff, blackish bristles. The bright red berries occur

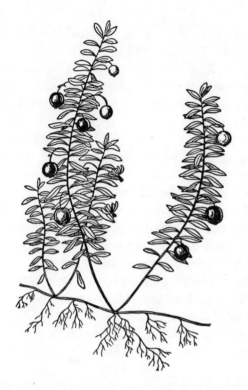

from fall throughout winter in some areas, growing in terminal clusters like the blueberries. Their taste is sharply acidic.

The mountain-cranberry is well known to northerners. It is a popular wild fruit, and at times is gathered and sold on the markets of the big cities. The berries are deemed superior in taste after they have been frosted at least once. An inhabitant of dry, acid soils, the plant is common in the barrenlands of subarctic America. Its range stretches across the northland from east to west, and as far south as New England.

The remaining two species of cranberries, except for leaf shape and size of fruit, look essentially alike. The small bog cranberry, *Vaccinium oxycoccus*, has small, triangular-shaped leaves with white undersides. Its berries are also small, and usually speckled or somewhat mottled with a drabness. A resident of boggy, acidic soil, this northerner is common from Labrador to Alaska, and southward to the Great Lakes, New England, and the Virginias.

The large or American bog cranberry, *Vaccinium macrocarpon*, is the common cultivated cranberry. Its flavor is still superior, though, when harvested from the wilds. The leaves of this species are oblong-shaped and slightly whitish beneath. The berries are thick and run up to ½ inch in diameter. In areas where the plants are abundant, it does not take long to gather a mess of fruit big enough to keep you in cranberry sauce all year. The large bog cranberry prefers open, boggy areas, as well as wet shores and fairly heavy swamp areas. This species is more southerly, ranging from Newfoundland to Wisconsin and south to the Carolinas. The fruit of both bog cranberries clings to its evergreen bushes long into the winter, through many frosts.

Cranberries must be cleaned of all their stem and leaf bits before they can be prepared for the table. They are edible raw, but not many people cotton to their sour taste. Raw cranberries can be stored in a refrigerator or dry cold storage for months in a plastic bag with a few air holes provided. They can also be frozen raw for short periods of time. Once cooked, they can be canned or frozen with ease.

Our simplest whole-berry cranberry sauce is made by dissolving 2 cups sugar in 2½ to 3 cups water. Then add 6 cups berries and simmer for 15 minutes until the berries have cooked and popped their skins, and the mixture begins to thicken. For a spicier sauce, add 1 apple quartered, at the beginning of cooking, 2 tablespoons orange juice, ½ cup slivered almonds, ½ cup raisins, and a dash of freshly ground ginger.

If you are not fond of whole-berry sauces, try cranberry jelly. This is simple. Stir 5 cups of cranberries into 2 cups of boiling water. Then cook for 20 minutes, rub through a sieve, and cook another 5 minutes. Now add 2 cups sugar, and a dash each of salt, cinnamon, nutmeg, and allspice or ginger. Cook for an additional 2 minutes to blend the flavors.

In our house, cranberries are not restricted to the dinner table at turkey time. They are eaten almost the year around—as a condiment, in salads, and for dessert.

For a delightful molded cranberry salad, soak 1 tablespoon of unflavored gelatin in 3 tablespoons water, and set aside. Meanwhile, cook 2 cups cranberries in 1 cup boiling water for about 15 minutes, then strain. Add ½ cup sugar, a dash of salt, and the softened gelatin. Now chill. When the gelatin begins to set, fold in ½ cup diced celery, ½ cup crushed pineapple, ½ cup chopped walnuts, and ½ cup cottage cheese. Pour into a greased mold, and chill until set.

HIGH-BUSH CRANBERRY
Viburnum spp.

The landscape of northern Ontario in the dead of winter is always made brighter by the sparkling, snow-covered, red berries of the high-bush cranberry. Long after the shrub has lost all of its leaves, the bright, still-hanging berries add a splash of color to the otherwise snow-whitened poplar and birch woods. The frozen berries make interesting nibbling if you are out snowshoeing or on cross-country skis. Even the ruffed grouse enjoy their sweet-sour flavor.

There are several species of high-bush cranberries in North America, none of which are true cranberries. They are rather members of the genus *Viburnum,* of the honeysuckle family. The common high-bush cranberry, *Viburnum trilobum,* is a large, coarse shrub growing to twelve or so feet in height and sporting grey-brown bark. As the species name *trilobum* suggests, the leaves of this shrub are three-lobed towards the outside. They somewhat resemble the leaves of the maple, especially late in the year when they change their coat of green for one of bright autumnal color.

Long branches bearing flat-topped flower clusters in late spring turn to supporting clusters of red to orange fruit in late September and throughout winter. The fruit is acidic, juicy, and contains a single flat stone per fruit. Because of this stone, many people preparing high-bush cranberries for the table prefer to strain the preserves before serving.

The common high-bush cranberry is f o u n d from Newfoundland to British Columbia, s o u t h through eastern Canada to New England, and locally through the Midwest. It is an inhabitant of rich, cool woods, stream and brook banks, rocky slopes, and boundaries such as fencerows and old stump or stone walls.

Another similar wild edible is the mooseberry or squashberry (*V i b u r n u m edule*), a somewhat straggly looking, grey - barked shrub whose fruit is also collected for making jams and jellies. This plant's range is a little broader, also extending farther north, and accordingly its reddish-orange fruit ripens a little earlier. The fruit of the high-bush cranberries is most often used in jellies.

These are very simple to make by combining berries and water in a saucepan in proportions of 2 parts fruit to 3 parts liquid. Bring to a boil and simmer until the fruit is soft. Then put the cooked fruit through a fine strainer or a jelly bag. Add ⅔ to ¾ cup of sugar (and a dash of cinnamon, if you like) to each cup of fruit, and again slowly bring to a boil. This jelly will keep in the refrigerator for some time, or sealed in glass canning jars, will last all year.

The sweet-sour taste of the high-bush cranberry is not something that everyone catches onto on the first try. We have introduced many people to these high-bush berries in a molded cranberry salad. More often than not, they become converted.

For another treat, combine 2 cups of cranberries and an equal amount of cooking apples cut into small pieces, with ¾ cup water in a saucepan. Bring to a slow bubble, and cook until the fruit is soft and mushy. Put through a strainer, and to the purée add 1 cup of sugar, a dash of allspice, and a teaspoon of grated lemon and orange rinds. This wild "cranapple" sauce really adds zip to your favorite pork dish.

RASPBERRY & BLACKBERRY
Rubus spp.

Raspberry

There are so many raspberries and blackberries in North America that it would take a week just to read descriptions of them in a text on taxonomic botany. Some botanists conservatively estimate two hundred different species; others go to over twice that. The g e n u s *Rubus,* of the rose family, is quite widespread across this continent, and encompasses the raspberries, blackberries, dewberries and cloud-

Blackberry

berries. Of course, a group of plants this large would promote a variety of local common names. Among the more popular are thimbleberries, blackcaps, baked-apple berries, plumboys, salmonberries, and heaps of others.

But enough of the refinements of correctly naming what you are eating. The wild harvester need only know that the raspberries and blackberries are mostly shrubs, some creeping and some tall and upright, some with spines and some without. The flowering stems tend to become woody with age, especially near the base. The alternate leaves range from simple to compound, and are sometimes palmately lobed or toothed. The fruits are all variations on the same theme. All are compound, that is made up of many tiny fruits each containing one hard seed. All resemble the cultivated fruits of the marketplace; however, a lot of variation in size and fruit color occurs in the wild species. The "berries" range from red to purple, black, deep blue, and even yellow and orange. All are juicy, tasty, and high in nutritive value.

The various *Rubus* members are found from the Arctic to Oklahoma, Virginia and even Georgia, from the east coast to the west, on fairly low to mountainous terrain, in rich and poor soils, along fencerows, in clearings, in open woods, and in thick brush.

Both the black raspberry and various types of blackberries were used long ago by Indian tribes as a remedy for dysentery. The fruits also constituted a part of the natives' diet.

For years, wild raspberries and blackberries have constituted an important part of our diet as well. We first get our fill of the delicious fruits by eating them raw with a bit of sugar and cream, or with dry cereals. We also freeze these wild fruits quite successfully in a thin, sugary syrup. We enjoy both jams and jellies made with these wild edibles, and frequently combine them with currants, gooseberries, cherries, strawberries, or other fruits to make delicious preserves that will keep all year once sealed in sterilized canning jars.

For the simplest wild berry jam, heat 4 cups of crushed raspberries or blackberries (or the two combined) in a large kettle. Then add 3 cups of sugar, stirring constantly and cooking until the jam is thick and jelly-like.

Wild raspberries and blackberries also make delicious baked desserts. A deep-dish wild berry pie is hard to beat, baked slowly with only a little flour and sugar added to the berries, and a tender, flaky upper crust. Or, if you prefer the rich, heavy pies, try this one. Sprinkle ¾ cup of sugar over 2 cups cleaned, packed raspberries or blackberries. Let this stand long enough for the fruits to become thoroughly sweetened. Meanwhile, in a saucepan, mix 1½ tablespoons unflavored gelatin, ⅓ cup sugar, ¾ cup water, 1 tablespoon lemon juice, and a dash of salt. Stir constantly over low heat until the gelatin dissolves. Add the berry mixture, and chill until the gelatin begins to set. Then fold in 2 well beaten egg whites and 1 cup of whipped cream. Pour into a baked graham cracker shell, decorate with whole berries, and chill until well set.

For another dessert favorite, combine 1½ cups flour, ½ cup sugar, 1 tablespoon baking powder, and ½ teaspoon salt in a bowl. In a separate bowl, beat 1 egg, ½ cup cereal cream, and ½ cup melted butter. Gently stir the dry ingredients into the cream and egg mixture, and pour the batter into a square, greased baking dish. Bake in a hot oven for about 25 minutes, until golden and crusty. Cut into squares and serve warm with brandied wild raspberry sauce liberally poured over the top.

For the sauce, stir 2 tablespoons of cornstarch into 2 tablespoons of cold water. Then, to 2 cups of freshly crushed, packed raspberries, add 2 cups sugar, ½ cup melted butter, and the cornstarch paste. Bring to a slow boil, stirring constantly, and simmer for a couple of minutes, until thickened. Then remove from heat and add 1 teaspoon of lemon juice and ¼ to ½ cup of brandy, depending on your tastes. This sauce is delightful either warm or cooled.

The raspberries and blackberries, alone or in combination with other fruits, also make some of the best wines and cordials I have ever tasted. Wild black raspberry cordial has long been my favorite. Raspberry wine has a full-bodied fruity flavor that is hard not to enjoy. And if you really want an after-dinner drink that finishes off a hearty meal with perfection, try blackberry juice and brandy, half and half, swirled in a warmed brandy snifter.

STRAWBERRY

Fragaria spp.

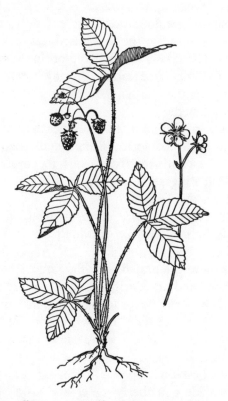

One of the wild fruits most easily recognized when seen for the first time is the wild strawberry. These miniature models of the cultivated types are among the most popular of the edible wild fruits, with both two- and four-legged creatures alike, and often grow in amazing abundance during early summer.

As a child, I can well remember wild strawberry season as a time of warm, sunny afternoons s p e n t roaming the nearby fields collecting the sweet, juicy fruits that made the best strawberry shortcake of the year. But the outcome of these afternoons bore other after-effects as well. Come evening, a slightly ill-at-ease feeling ran through the children, whose stomachs had been the recipients of too many handfuls of berries that never quite made it as far as the basket or pail. There was also a rash of hot baths and "sundowners" for the adults, who often suffered aches and stiffness resultant of an afternoon of unaccustomed bending and crouching to harvest the low-lying plants. The aches and pains, along with the following morning's task of cleaning, cooking, and canning, may have seemed to make the fruit not worth the trouble it cost, but in the dead of winter when the wild strawberry jam was served with hot tea biscuits and butter, the collecting and kitchen preparations were looked upon in a different light, and always with the resolution to gather more next year.

Several species of strawberries exist in North America. They are all basically similar in appearance, and hardly need any description. Almost anyone who has ever walked through a country field or meadow in early summer has found and tasted the small, sweet fruits. The common wild or scarlet strawberry, *Fragaria virginiana,* is probably the most widely harvested. A perennial member of the rose family, the leaves arise on long leaf stalks directly from the roots. Both leaves and stalks are covered with soft hairs. The leaves are made up of three small leaflets, each sharply toothed. The plants spread readily by means of runners. White flowers give rise to the pulpy, red fruits, which are actually a multitude of fruits combined.

The common wild strawberry can be found from southeastern Canada to the Prairie Provinces, and south to Florida, Texas, and Arizona. Other species together inhabit the continent from coast to coast, in pastures, meadows, open fields, clearings, dry hillsides and slopes, and almost every wild place that is not overly arid.

The European woodland strawberry, *Fragaria vesca,* is a husky, robust version of the common wild berry. Many people don't consider its flavor to be quite as good.

What wild strawberry dish is more celebrated than the famous berry shortcake? And because of its fame and popularity, hundreds of recipes for this dessert delicacy exist. One of our favorites is a campfire dish, using bannock or frying pan bread as the base. But when you are not in camp and have the convenience of your kitchen oven at hand, try it with a rich, moist sponge cake and cream.

Wild strawberry season is the best time of year to get out the old ice cream freezer, because homemade strawberry ice cream has a taste like nothing you buy in your local supermarket. The recipe is simple. Just crush 1 quart of cleaned strawberries and sprinkle with 1 cup of sugar. Mix the fruit and sugar thoroughly, then chill. Add 4 cups of cream. Pack and freeze according to the directions on your ice cream freezer.

For anyone without a home ice cream maker, try crushing and cooking 1½ pints strawberries over low heat with 1 cup sugar. Chill completely and add 2 tablespoons lemon juice. Next, whip 1 cup of whipping cream. Separately, beat 2 egg whites with a dash of salt and cream of tartar. Fold both whipped ingredients into the chilled berries. Then pour into a mold or just into ice cube trays, and freeze until solid.

Strawberries make delicious pies. Simply filling a deep pie dish with the wild berries, sprinkling with a little flour and sugar, and then baking with only a top crust, will produce a delicious wild berry feast. Our favorite two-crust strawberry and rhubarb pie also tastes best when prepared with wild berries. Or try strawberries and bananas together as a pie filling. The possibilities are unlimited.

Strawberry tea has long been a favorite among outdoorsmen. Made simply by steeping strawberry leaves in boiling water, its taste is quite refreshing. And, of course, anyone who has never had the opportunity of sipping strawberry wine is really missing something!

Who could forget wild strawberry preserves? For this, combine 2 cups berries and 2 cups sugar in a large kettle and let stand for an hour. Bring to a rolling boil, and stir constantly for 10 minutes. Add 1 tablespoon of lemon juice if desired, scoop the white foam from the top of the kettle, and seal immediately in glass canning jars.

GOOSEBERRY
Ribes spp.

Wild fruits comprised a significant part of the diets of many Indians and early settlers. Many were eaten fresh, while others were dried in the sun for future use in breads and puddings. These fruits frequently provided a good source of the sunshine vitamin for the Indians and colonists. Among the fruits that were harvested for the pot were the gooseberries — plump, juicy members of the saxifrage family which occasionally sport bristly or prickly fruits. But the clever natives knew how to get past the spines. They are reported to have rid the fruit of their bristles by singeing them over hot coals.

Some ten to twenty species of gooseberries reside on this continent, from Newfoundland to the West Coast, and south below the Mason-Dixon line. The bushes generally grow to about three to four feet in height, with slender, graceful and sometimes drooping branches. The alternate leaves are generally three- to five-lobed, and somewhat rounded or heart-shaped in outline. White to greenish flowers give rise to deep brownish-red to purplish fruits with thin skins. The fruits of the prickly gooseberry, *Ribes cynosbati,* are armed with long bristles, although occasionally nearly smooth specimens are found. The fruits of

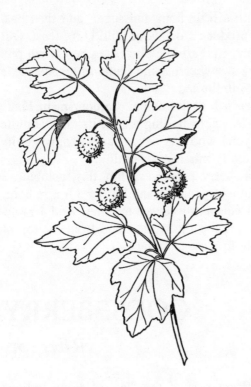

the smooth northern gooseberry, *Ribes oxyacanthoides*, are, as the name indicates, without prickles.

The gooseberries generally favor rocky woods and open, loamy or sometimes swampy areas, but local species can be found in m a n y varying habitats. Quite a number of European gooseberries, the cultivated species, have also escaped their plantings and established themselves in the wilds.

Gooseberries are relatively easy to clean. Simply remove the bits of stems and wash. We enjoy them a number of ways, but with Indian curries we especially like fresh gooseberry chutney. For this delightful relish, chop 5 cups of berries, 2 cups of seedless raisins and 2 sweet onions in a food grinder. Then add 1 cup brown sugar, 3 tablespoons each of ginger and dry mustard, 1 tablespoon salt, ¼ teaspoon cayenne, ½ teaspoon turmeric, ½ teaspoon ground coriander, and 4 cups mild vinegar. Bring the mixture to the boiling point and simmer for about an hour. Some people prefer to strain this chutney before serving, but we like ours with the chunks and seeds.

If you would prefer gooseberries for dessert, try them in a fresh gooseberry pie. For the filling that goes between the double crusts, combine 1 cup sugar, ¼ cup flour, 1 tablespoon quick-cooking tapioca, 2 tablespoons lemon juice, and ½ teaspoon each of cinnamon and mace. Sprinkle this over 4 cups of gooseberries and mix. Before putting on the top crust, dot liberally with butter.

CURRANT

Ribes spp.

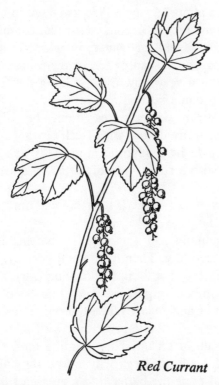

Red Currant

Dried wild currants were probably a staple source of vitamin C for the Indians long before the first white men ever set foot on our eastern shores. In fact, the Indians are reported to have eaten other parts of the plant, including the young leaves and flowers, as well as the berries. When the early settlers arrived, they too utilized the berries — raw, cooked, and dried.

Very close relatives of the gooseberries, there is a strong resemblance between the two groups of plants. The currants are mostly low bushes, under five feet in height. The slender, graceful branches often recline, and sometimes even touch the ground, taking root where they touch. Some branches, though, are upright. The leaves are three- to five-lobed. Clusters of flowers vary in color with species, from off-white to yellowish and greenish and purplish-brown. Basically two types of berries occur — red and black — although one form has yellow fruit. The berries are mostly shiny, firm, and juicy, with few or no bristles.

Numerous species of currants abound across the whole of North America. Generally they require some moisture, but are found in a variety of habitats. In the mountainous regions of the West, they take to shaded or open land where there is plenty of soil moisture. The plants seek out areas near springs and gullies in desert areas as well, where the source of water isn't too far underground. The common red currant, which thrives in the East, prefers thickets, cool woods, and even swamps. The wild black currant likes rich slopes and thickets. The various species of currants occur east to west from Labrador to the Mackenzie River valley and the West Coast, and south as far as North Carolina and Missouri.

In our household, currants are almost a staple. We use them in chutneys and relishes, in cakes and muffins, in breads and rolls, combined with raisins in many dessert dishes, and, of course, in wine. To preserve the fruits for use out-of-season, we can or freeze them in a light sugar syrup. We also dry some in a very slow oven, mixing the fruits often to ensure that they don't become burned or over-dried.

One of the most delightful currant desserts I can remember was a heavy, homemade pound cake liberally dotted with fresh currants. Another memorable dessert was made by a favorite aunt. This delicacy was a fresh currant pie topped with a spicy, crumbly crust, and served warm with home-frozen ice cream.

But one of the best currant desserts, we feel, is what we call our "currant snacking cake." For this cake, cream ½ cup butter with ½ teaspoon pure vanilla extract. Gradually add 1 cup sugar, beating in a little at a time to keep the mixture fluffy. Then, while still beating, add 2 egg yolks. Now, separately, mix together 2 cups cake and pastry flour, ½ teaspoon salt, and 1 tablespoon baking powder. Add about ¼ of the flour mixture to the creamed butter and sugar. When thoroughly mixed, add ¼ cup of cereal cream. Repeat the same procedure with a second ¼ of the flour and another ¼ cup of cream. Then add the remainder of the flour and beat well. Fold 2 whipped egg whites into the batter. Then fold 1 cup currants in, and pour the batter into a greased square baking pan. It will take ½ to ¾ of an hour in a moderately hot oven for the baking to finish. Then cool on a wire rack before cutting. This cake makes terrific lunch-box fare.

Currants, of course, make some of the best jams and jellies you have ever tasted. And wild currant and sour cherry wine has a uniquely fruity flavor.

ELDERBERRY

Sambucus spp.

Almost everyone who has ever heard of elderberries has heard of elderberry wine. It is almost a part of American heritage. As far back as 1820, elderberry juice was recognized in the *U. S. Pharmacopoeia* as a prime wine ingredient. And if you have ever had the opportunity to sip good elderberry wine, you know why it was so popular.

The plant was also used as a cure for a number of disorders, including colic, headache, constipation, and in treating wounds and inflammations. The flowers of the elderberry, also known as the common, American or sweet elder, reputedly induced sweating and could be used to reduce fevers. About a half dozen species of elder grow in North America. The common elder, *Sambucus canadensis,* is perhaps the best known. Reaching six to twelve feet in height and higher in southern climates, the elder is a stout shrub. Its thick stems are generally upright and often grow in clusters. The young stems are characterized by a greenish color and very little bark. A cross-section of such stems reveals a large, white, pithy area. As the shrub matures, the bark becomes greyish-brown, the wood thickens, and the pith shrinks. Five to eleven leaflets make up each leaf, the lower ones often lobed. The leaves arise opposite one another on the stem.

In June and July, relatively broad and flat clusters of fragrant whitish flowers appear. By late summer or early fall, these are replaced by round, juicy, purplish-black berries, each usually containing three or four seeds. The acid content of the berries is not high, hence they are not sour.

The common elder is a resident of fencerows, roadsides, old building walls, embankments, ditches, and many other moist, fertile soil areas. Its range extends from northern Nova Scotia to Manitoba, and south to the Gulf Coast and Oklahoma.

The red-berried elder is also quite common on this continent. However, the berries are not the most pleasant in flavor. Some even claim them to cause stomach disorders in sensitive individuals; others claim they are poisonous.

The blue-berried elder is a large plant common in the West. It produces large, sweet, juicy fruits that are popular for making jelly.

Some people distinctly dislike the flavor of fresh wild elderberries. Because of their low acid content, they are not particularly tasty raw; but cooked is a different story. Stewed with a suitable amount of sugar (not too much, depending on your taste) and large chunks of lemon rind, the berries can be canned or frozen for year-round use. An easier and perhaps better way to deal with them is to spread the cleaned berries out on newspapers and dry them in a warm, dry place. Once they are dry, they can be packed away for storage. To use them, stew them with just a little water, sugar to taste, and lemon rind.

For a hearty jelly to serve with hot muffins or tea biscuits at Sunday morning brunch, stew 3 quarts of elderberries and 3 cups peeled, sliced apples in as little water as possible. In a separate pan, stew 3 quarts gooseberries, again in as little water as possible. Then combine the cooked fruits, add the juice of ½ orange and ½ lemon, and squeeze through a jelly bag. For each cup of liquid gleaned, add 1 cup of sugar. Then follow your favorite recipe for boiling and testing the jelly. When the jelly drips from a spoon in a solid sheet, it is ready for the canning and sealing operation. You should not cook the liquid to the stage where the jelly sheets if you would prefer a softer consistency. Once the sheeting stage has been reached, you are assured of a fairly firm jelly.

We have always been fond of elderberry wine. But we also like an elderberry after-dinner drink that we call "elderberry pickling juice." To make this, pack 2 quarts of elderberries and a touch of mace and cinnamon into a glass crock. Cover with 2 quarts of a not-too-expensive brandy, and let the mixture sit for at least six weeks before sampling. The resultant liquid is the perfect wind-up for a Canada goose and wild rice dinner. The fruit can also be sampled in small quantities, but it is pretty heady.

SERVICEBERRY
Amelanchier spp.

The serviceberries, like many other species bearing common misnomers, do not produce berries. Their fruits may resemble berries, but they are

actually small pomes, the same type of fruit as the commonplace apple. Each of these pomes is composed of ten compartments housing one seed. Thus, when all ripen, the serviceberries can be pretty seedy. But their seeds have somewhat of a nutty taste, not at all unpleasant, and lend a little flavor to the otherwise bland fruit.

The serviceberries are not highly acidic, hence their flavor varies between bland and slightly sweet. They are not terribly appetizing when raw, but are excellent dried and quite good once stewed with sugar and lemon juice.

The serviceberries bear a host of common names, among which are juneberry, shadberry, shadbush, sarviceberry, sugarplum, sugarpear, Indian pear, and Saskatoonberry. There are over a score of serviceberry species on this continent. They range as far north as southern Labrador and Alaska, and flourish southwards, east and west, to California and the Gulf Coast. The serviceberries generally prefer open country, rocky or gravelly areas such as slopes or banks, and edges of woods and swamps. In the south, the fruit can be harvested as early as May, but farther north you have to wait until mid- to late summer.

The serviceberries are large shrubs or small trees. The single leaves arise alternately on slender leaf stalks. The margins of the leaves vary from almost smooth to very coarsely toothed. Long, white, drooping flower clusters adorn most species, but some are single-flowered. The attractive flowers mature to form round or slightly pear-shaped fruits, bluish-black or purplish in color, with a sweet, pulpy flesh.

We prefer to dry serviceberries after we have collected them. (The attic or the oven are both okay for this.) In fact, we used to get an annual Christmas shipment of dried berries from friends who lived closer to an abundant supply than we did. Those shipments were most welcome, but unfortunately ended when our suppliers moved to the East Coast.

The Indians supposedly used dried serviceberries, much pounded and beaten, in making bread. We prefer to use ours in serviceberry cupcakes. Combine 2 cups flour, 1 teaspoon bicarbonate of soda, ½ teaspoon salt, and ½ teaspoon each of cinnamon, cloves, nutmeg, and mace in a bowl. In another bowl, beat ½ cup butter, 1 cup firmly packed brown sugar, and 1 egg until thoroughly blended and fluffy. Now, add the contents of bowl 1, a third at a time, to those of bowl 2, alternately beating in a total of 1 cup of milk. It is important not to over-mix. Finally, add 1 cup dried serviceberries, ½ cup dried currants, and 1 cup chopped walnuts. Pour the thick batter into greased cupcake cups and bake in a moderately hot oven until done. After cooling on a wire rack, you can frost the cupcakes with a creamy rum frosting, or eat them just as they are.

If you want to add zip to your breakfast waffles, try this. For your waffle batter, beat ½ cup sour cream with 1 cup milk, 2 eggs, and ¼ cup vegetable oil. Then blend in a mixture of 1½ cups flour, 1 tablespoon baking powder, ½ teaspoon baking soda, 1 tablespoon sugar, and ½ teaspoon salt. The batter need not be smooth. If the batter appears too thick, add more milk. Now mix in 1 cup dried serviceberries, and ½ cup of chopped pecans. With pats of butter and loads of maple syrup, these will be the best waffles you will ever remember.

MULBERRY

Morus spp.

Another group of north temperate zone plants that provide plenty of fruit that is free-for-the-taking is the mulberry. When early summer comes and the fruit ripens, it can often be collected in amazing abundance. Where the plants are plentiful, one afternoon's outing can usually produce enough mulberry preserves to last until next year's crop.

The red mulberry, *Morus rubra*, probably makes the best eating of the ten or more species that exist on this continent. It is a small tree, averaging twenty to thirty feet in height. The trunk is usually quite thick and the top spreading and crookedly branching. Irregularly shaped leaves, rough on the upper surface and softly pubescent beneath, arise alternately and grow from three to five inches in length.

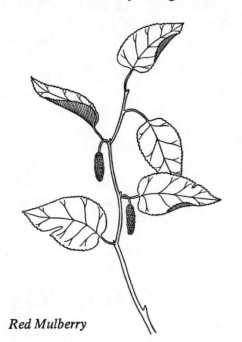

Red Mulberry

The leaves are a dark green in color and have pointed tips and sharply toothed edges. As with all members of the mulberry family, the stems house a milky juice.

In the axils of the leaves arise greenish spikes of flowers. Male and female develop on different spikes, but often on the same tree. The spikes of flowers give rise to long clusters of fruits, which are bright red when they first ripen and turn to dark purple as they become older. The dark fruits look not unlike those of the blackberry.

The red mulberry is a lover of rich soils. It grows from southern Ontario to Minnesota and South Dakota, and from New England to Texas and Florida.

The white mulberry, *Morus alba*, best known in past generations as food for silkworms, has spread from cultivation in some parts of the east. Its fruit is whitish to pinkish, but not as juicy as that of the native red species.

The small Texas mulberry, *Morus microphylla,* is a native of this continent, and is popular in the western states to northern Mexico as a wild edible.

The mulberries can be eaten raw, but generally reach their peak flavor when cooked. However, cleaned, sprinkled with a little sugar, and covered with fresh table cream, they are well worth trying.

Mulberry jelly is not to be overlooked. For a bright, firm jelly, use 2 cups of immature red berries and 2 cups of mature, purple berries, both mashed. Add ¼ cup of water and slowly bring to a boil. Boil for 10 minutes and then strain through a fine cloth. To the juice, add 1 tablespoon lemon juice, 2 cups sugar, and 1 package powdered or liquid pectin. Continue boiling until you reach the jelly consistency you would like. Then pour into sterilized glass jars and seal.

Mulberries make one of our favorite fillings for a jam roll. To begin, mix together 2 cups flour, 1½ tablespoons baking powder, 1 teaspoon salt, and 1 tablespoon sugar. With a pastry blender, cut 6 tablespoons butter into the flour mixture. Then add ¾ cup of cereal cream. When the dough is completely mixed, turn it onto a floured board or pastry sheet and knead for a few seconds. Then roll the pastry out until it is about ⅓ inch thick. Brush melted butter over it with a pastry brush. Sprinkle 3 cups of ripe mulberries over the dough, along with ½ cup of brown sugar (packed), and a dash of cinnamon and nutmeg. Now roll the dough up like a jelly roll and place it with the seam down in a greased baking dish. Brush with milk and beaten egg and bake in a hot oven for about ½ hour. Then cool slightly. Slice and serve warm with brandied hard sauce and stewed mulberries. When mulberries are out of season, this delightful roll can be made with mulberry preserves.

WILD ROSE
Rosa spp.

To many, "rose" conjures up the image of the long-stemmed beauty of St. Valentine's Day and the thorny, colorful bushes of the back

garden. To the wild harvester, it means jellies, teas, jams, sauces, and purées that all have a special flavor.

Everyone knows what a rose shrub looks like. Its alternate, five-leaflet leaves have finely toothed edges and are a rich green in color. The branches often form dense thickets, almost impenetrable because of the presence of numerous briars and thorns on the stems and stalks. Solitary, fragrant, five-petalled flowers vary from pinkish-white to deep rose in color, and give rise to the fruit, which is known as a "hip" or a "haw." These hips are somewhat rounded, smooth, and narrowed at one end. They are delicate in taste, something like a mild apple in flavor, and house a central core of seeds. The fruits hang on to the bushes long after the ground has been frozen, and oftentimes last through most of the winter.

Not only do rose hips taste good, but they are good for you. Their vitamin C content is so high that they were feverishly gathered in Europe during the last war to nourish the fighting troops. It is said that rose hip juice has twenty-four times more vitamin C per unit than orange juice. A couple of these small, pulpy fruits per day would surely keep cold sufferers happier!

Roses hybridize so readily that it is almost impossible to guess how many different species exist on this continent. Most botanists guess about thirty-five or more. They are all inhabitants of moist areas such as open woods and fields, stream banks and meadows, and are common along roadsides and fencerows. They are found quite far north, and can tolerate extremely cold winter temperatures. Some people claim the hips are not at their best until they have been frosted at least three times.

There are so many different things than can be done with rose hips that it is difficult to know which ones to mention. These fruits are tasty when eaten fresh, and equally good when dried and used for snacking. Rose hips are thoroughly enjoyable in a Waldorf-type salad made by tossing together 1 cup diced celery, ½ cup diced apples, 1 cup diced rose hips, ½ cup chopped walnuts, and ½ cup mayonnaise with lemon.

For the best applesauce you have ever tasted, pare and core good saucing apples and toss them into a pressure cooker with ¼ to ½ cup of water. Add a touch of salt, and cook at 15 pounds pressure for 5 minutes. (Without a pressure cooker, simply stew over low heat until tender.) Then put the cooked apples through a strainer, or leave them lumpy as we do. Add a bit of sugar to taste and cool. For every 5 cups of applesauce, add 1 cup of rose hip syrup made by boiling 2 cups of rose hips in 1 cup water until soft. Then strain the rose hips through a jelly bag and add 1 cup sugar. Boil this syrup for 5 minutes. Rose hip syrup adds a lot of zip to applesauce, and this combination is a real pick-up for pork, bear, or venison.

For a deliciously different jam, simmer 6 cups rose hips in 2 cups water until soft (about ½ hour). Then press through a sieve to make a purée. To the purée, add the juice of one lemon and 3 cups of sugar. Cook until the mixture is thick, and then bottle in sterilized sealers.

HAWTHORN
Crataegus spp.

Along the back of my parents' property line grows an impenetrable tangle of hawthorns. They are on the neighbor's side of the fence, and every year are subjected to continuous threats with the chain saw and ax because their thickness blocks much of the light coming into the back yard. But the hawthorn jungle has its advantages. It provides welcome shade and privacy on hot summer days. Its mass of sharp thorns keeps the neighborhood dogs and stray kids away. The shrubs house all manner of interesting winged creatures. And the fruit makes great jelly.

The hawthorn, or haw as it is somewhat more commonly called, is a relative of the rose and the apple. It bears a host of other common

names, including thorn, thornplum, thornapple, and cockspur thorn. The genus name *Crataegus* comes from the Greek word *kratos*, meaning strength. And the genus is a large one. Certainly over a hundred species exist on this continent, and perhaps three or four times that many are distinguished.

The hawthorns are all shrubs or small trees. They are characterized by long, sharp spines or thorns and simple, alternate, lobed leaves. In spring, they put on a showy display of five-petalled whitish flowers. These grow in terminal clusters, and give rise to numerous small, red, yellow, or deep purple pome-type fruits. Each fruit contains up to five nut-like seeds surrounded with pulpy and sometimes dry flesh. The fruit remains on the plant long after the snow has arrived.

The hawthorns are quite common in the central and northeastern portions of this continent. But they do grow from east to west and north to south. They are resident along stream and river banks, in woods, thickets, pastures and open fields, and along roadsides and fencerows.

All of the hawthorns are edible, but some are so lacking in flesh that they are hardly worth the trouble of collecting. Others are fleshy, but they don't taste very good, especially raw. You have to keep sampling until you come up with some really good specimens. At any rate, they almost all make good jams, jellies, and marmalade.

For a great jelly, crush 2 to 3 pounds of haws, add enough water to cover the fruit, and simmer for about 15 minutes. Then strain through a jelly bag. For each 4 cups of juice, add 1 package of pectin (or, if you are using not-too-ripe fruit, you can probably omit the pectin) and 6 to 7 cups of sugar. Bring to a rolling boil and cook for 2 minutes or until the jelly sheets. Then skim off the foam and bottle in sterilized glass sealers.

PAWPAW
Asimina triloba

The pawpaw (sometimes spelled without the first "w" — papaw) has always inspired differences of opinion as far as edibility and choiceness are concerned. Some love the rich, custardy fruits; others dislike their taste immensely. I must admit not having thought much of the pawpaw the first time I sampled it, but it grows on you. Sooner or later, everyone becomes converted.

The pawpaw, or custard apple or false banana as it is sometimes called, is a northern member of an otherwise tropical group of plants. It frequents the sunny south from Florida to Texas, but also grows as far north as New York and New Jersey, Michigan and Iowa. The pawpaw can also be found along the northern shore of Lake Erie and Lake St. Clair in southern Ontario, where we usually make an annual sojourn, after the first frost, to collect the ripe fruits.

The custard apple is a large shrub or small tree, generally under forty feet in height and about six inches in trunk diameter. This resident of rich, moist, fertile woods looks like it belongs in the tropics. Its leaves are covered with a rusty-colored fuzz when very young, but soon expand to form dark green, somewhat drooping ovals with tapered bases, pointed tips, and smooth margins. The leaves run up to ten inches in length. The flowers appear in the axils of the previous year's alternate vegetative growth. These are quite large, with two whorls of three petals each. The outer petals are large; the inner ones considerably smaller. They are conspicuously veined, greenish when young, and turning to a deep purplish later.

The fruit of the custard apple is most interesting. It vaguely resembles a short, thick banana, up to five inches in length. When ripe and still attached to the tree, it is yellowish, quite hard, and somewhat bitter. However, a few days after falling or being picked from the tree, it turns a rich brown and becomes soft and very sweet. Its pulp is a bright yellow, and surrounds large brown seeds that are not difficult to get rid of.

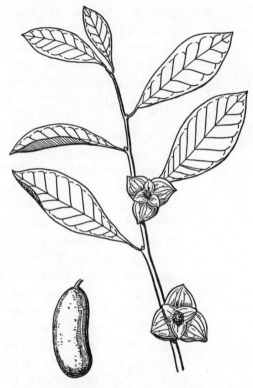

The ripe fruit has quite a pleasant fragrance.

The taste and texture of pawpaw makes it a natural for combining with eggs and milk. Try this for a delightful dessert. It is a recipe that has been going strong since the turn of the century. To 3 beaten egg yolks, add 1 cup whole milk, 1 cup cereal cream, and ½ cup sugar. Cook over low heat until hot and smooth. Then combine with 1 cup mashed, strained pawpaw pulp. Refrigerate, and just before serving top with meringue made by beating 3 egg whites with ¼ cup sugar and ½ teaspoon cream of tartar. Pop under the broiler just long enough for the meringue to brown, but make sure the pawpaw pudding remains cold.

MAY APPLE
Podophyllum peltatum

The May apple is a most interesting and attractive species. Although its fruit is edible and quite tasty, the remainder of the plant must be viewed with caution. The roots, stems, and leaves of this species are poisonous. However, in past history, these parts of the plant were used as remedies for various ailments by the natives of this continent. Its highly purgative nature made it a natural for a laxative, although I'm sure one had to watch the dosage quite carefully. Then the roots were used in another concoction to cure venereal disease, and one can see why. The

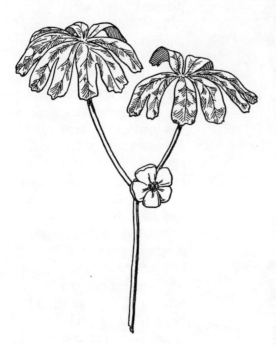

poisonous substances in the mixture probably promoted an illness in their own right, with a high enough body temperature to severely discourage any virus trying to survive in the system. It is even reported that some Indians used May apple plants as an insecticide on cultivated crops.

The May apple is a native North American. Perennial in nature, the plant spreads by means of long horizontal rootstocks. Each spring the roots give rise to two types of stems, a foot high or slightly taller. The ones with only one single, umbrella-shaped, five- to nine-lobed leaf, joined to the stalk at the center, do not bear flowers. These leaves are shaped somewhat like a shield and grow up to a foot in diameter. The second type of stem bears two leaves, each a bit smaller, and attached nearer the inner edge. These plants bear flowering stems.

The flowers of the May apple are most attractive, although unfortunately rather foul in odor. The almost two-inch wide flower is creamy white, waxy, and nods from the axils of the leaves. Between six and nine petals are common, with twice as many stamens. From the flowers develop long, yellow, egg-shaped fruits. The fruits are pulpy, seedy, and have a tough skin. The fruits are ready to be eaten in late summer, when the rest of the plant is dying.

A member of the barberry family, the May apple is also known as hog-apple, wild lemon, raccoonberry, and mandrake, although the latter is something of a misnomer because this species is not what is commonly recognized as the true mandrake. The May apple is found from western Quebec and southern Ontario, south to Florida, and west to Texas, Kansas, and Minnesota. Residents of rich soil areas, the plants are at times extremely plentiful in woods, thickets, and pastures.

Although the May apple can be eaten raw, the delicate stomach may find too many of these uncooked fruits to have somewhat deleterious effects. Cooked, the fruits can be enjoyed in greater quantity. If you want to sample the flavor of the uncooked fruit, try squeezing the juice from the fruit and mixing it with a bit of sugar and some fruit punch or tropical fruit juice. It is also good mixed with lemonade.

May apple preserves are worth trying. Clean and peel enough May apples to give you about 6 or 8 cups of mashed pulp. Simmer the pulp for about ½ hour, with as little water as possible, being careful not to let the fruit burn. Press the fruit through a coarse sieve. Then, for every 4 cups of juicy pulp, add 1 package of pectin, the juice of ½ lemon, and 4 cups of sugar. Bring to a rapid boil, skim, and store in sterilized glass sealers.

APPLE

Pyrus spp.

Apples have long been residents of this continent. They found their way across the ocean from Europe and Asia shortly after our Pilgrim fathers, and started to spread out as soon as they set foot on dry ground. It is said that their colonization was nudged along by the famous Johnny Appleseed who helped to spread the species from the East to the Midwest. Wild apples and wild crabapples now thrive in this temperate land. Most are escapees from cultivation and do not bear the fat, lush fruits of their domestic counterparts. But their fruit is often plentiful, easily harvested, and chock full of flavor.

Apples have a magic all of their own. They are so familiar and popular that even the staunchest city girl has been far enough away from the cement canyons to have seen an apple tree. And almost every block of bungalows in suburbia boasts at least a couple of crabapples.

The apples and crabapples are shrubs or relatively short trees, sometimes reaching twenty-five feet in height. Their branches are at times short and crooked, and sometimes bear spines. They are most memor-

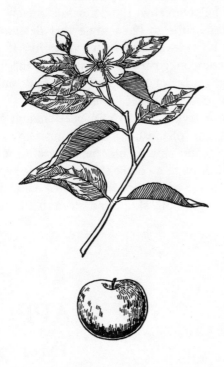

able in spring, when their whitish to pink, five-petalled blossoms flavor the countryside with an unforgettably delightful fragrance. T h e fruit, w h i c h is a fleshy pome, varies from an inch or less upwards in diameter. The flesh is tart, firm when freshly ripe, and delicious when cooked. Most wild apples are not terribly appetizing raw.

Ever since I was a kid, I have always loved applesauce. Like others, I enjoy it with pork. But I also like it with veal, venison, bear, beaver, some poultry dishes, as a dessert, for lunch, and even for breakfast. In short, I am an applesauce fanatic.

And there is nothing I like better than the tartness of applesauce made with wild apples. For fresh applesauce for breakfast, pare and core a half dozen apples when you are in the kitchen dealing with supper. Cut the apples into small pieces and put them in a stainless steel pot on the burner that vents your oven. Dissolve ½ cup of sugar and a dash of cinnamon in 1 cup of boiling water, and mix with the apples. Cover them and leave them on the warm burner through your cooking procedures, until the next morning. By breakfast time the apples will be deliciously soft, sweetened, and aged. A great start for a new day. And not too many calories, either!

Ever tried a wild apple omelet for breakfast? It is super. For the omelet, you will need a half dozen medium-sized apples. Pare and core the apples and cut them in six or eight pieces. Stew them until soft in enough water to keep the fruit from burning, adding just enough sugar to sweeten pleasantly. But don't add so much sugar as to kill the natural tartness. After the apples have been stewed and are slightly mushy, let them cool. Now crack 8 eggs into a bowl. Beat thoroughly with one cup of milk, and salt and pepper to taste. Pour the beaten eggs into a large, greased electric frying pan or omelet pan. Once the eggs are sufficiently cooked to fold easily, cover them with the applesauce mixture, and fold the omelet over the fruit. Put the top on the frying pan, and allow the omelet to cook slowly until it is heated through. Served with pure pork sausage and hot biscuits, this is an unexcelled Sunday breakfast.

There is just no end to what you can do with wild apples. Some of the old favorites include fried apples, apple pies and tarts, apple fritters, apple dumplings, apple pancackes and waffles, apple cake, and dozens of others. Here are a few recipes that we particularly enjoy.

One is for our own rich apple pudding. For the base, cut 1 cup of butter into 3 cups of flour. Add to this, 1 egg and ¼ cup water beaten together. Finally, add the juice of 1 lemon, a couple of spoonfuls of grated lemon rind and ⅓ cup of sugar. This will make a soft enough mixture to be pressed into a deep greased baking dish. But save some dough to decorate the top of the pudding in strips.

For the filling, mix together 4 cups peeled, cored and diced apples, 1 cup brown sugar, ½ teaspoon of cinnamon, ¼ teaspoon nutmeg, juice of one lemon, 1 cup seedless raisins, ½ cup currants, ½ cup blanched almond slivers, 1 beaten egg, and ¼ cup of evaporated milk. Place the filling in the crust and cover with strips of dough. Then pop into a moderately hot oven for about an hour. This heavy pudding is nice served warm with cream or sour cream.

And speaking of sour cream, here is one of the most delightful desserts you have ever tried. In a deep non-stick frying pan, melt 4 tablespoons butter and slowly pan-fry 5 cups of pared, cored, wild apple slices until the apples become soft. To the apples, add a previously mixed ½ cup sour cream, ¾ cup sugar, juice of 1 lemon, 2 tablespoons flour, 8 egg yolks, and ¾ cup of slivered almonds. Continue cooking slowly until this mixture begins to thicken. In the meantime, beat the 8 egg whites with ¼ teaspoon of salt until they have just reached stiffness. Fold the whites into the hot apple mixture, and then pour the mixture into a greased ovenproof casserole so that it is an inch or so deep. For the topping, mix ½ cup dry bread crumbs, ¼ cup white sugar, ¼ cup almond slivers, and ½ teaspoon cinnamon, and sprinkle over the oven-ready dessert. Bake in a moderate oven for about ¾ to 1 hour, or until firm. This dessert can be served either warm or cold, with whipped cream or sour cream.

Of course, you can't forget all of the old snacking standbys either, such as apple jelly, apple butter, or deep-fried apple rings. All are made just a touch more tart and tangy with wild apples.

RUM CHERRY
Prunus serotina

The rum c h e r r y , which gained its name some time back in the eastern history of this continent for its use as a mellowing a g e n t in rums and whiskies, is probably better known for its popularity among cabinet-makers. The wood of this cherry is very hard, and develops a deep, rich hue as it ages. Some of the most beautiful pieces of furniture I can remember w e r e the property of a Jesuit college, and were exquisitely crafted of this h a r d cherry. Rum cherry wood also makes outstanding gun stocks.

But the wood is not this species' only redeeming factor. The trees themselves are a delight to the eye, and their fruit makes delicious cherry treats. Commonly known as the wild black cherry, or simply the black cherry, this species is a native North American. It can be found from the Canadian Maritime Provinces, westward through southern Ontario to Minnesota and the Dakotas, and southward as far as Florida and Texas. It prefers dry woods and open areas, and is quite commonly found along fence-rows and old stone walls.

The rum cherry is a fair-sized tree, growing in some cases to ninety or more feet in height. The bark of its trunk is rough and dark, while that of the branches is smooth and a warm reddish-brown. When stripped, the inner bark is very aromatic, bearing the somewhat distinct odor of bitter almond. The leaves of the rum cherry are narrow, oblong and slightly lance-shaped, with a broader base and a pointed tip. Dark green and lustrous on the upper surface and paler beneath, the leaves sport bluntly toothed margins and a prominent midrib.

While the leaves are still expanding in the spring, numerous white flowers appear in long clusters. In late summer and early fall, grape-like clusters of rounded fruits, varying tremendously in size but at times three-quarters the size of domestic cherries, ripen. The fruits are a bright red at first, but darken with ripening to a deep purple or purplish-black. Eaten straight off the tree, their flavor is a little on the bitter-sweet side, but they are quite enjoyable. We find them delicious cooked.

Rum cherries are perhaps best known for their part in "cherry bounce," a lively concoction made by preserving the fruit in whiskey. As a child, I can remember my grandfather making this. Each Christmas we kids were allowed to sample a couple of cherries. I can still remember the ball of fire racing down my throat and into my stomach as I tasted my first cherry. I could not quite figure what all the adults saw in this burning liquid. Over the years, of course, my tastes have changed! Occasionally we make our own cherry bounce. It is made in a large, squat pottery crock by alternately layering cherries and sprinkling each layer with a spoonful of sugar mixed with a dash of cinnamon. When the crock is filled, we pour in Canadian rye whiskey to cover all of the fruit, and then seal it. For Christmas sampling, we usually start the brew just before the opening day of grouse season. In our neck of the woods, that is mid-September.

Rum cherries can be used in almost any cherry recipe, once suitably sweetened. They also make a pleasant jelly, especially if combined with apples. For an easy and tasty cherry dessert, clean and pit 4 cups of cherries. Put the cherries in a saucepan and pour over them ½ cup of boiling water. Slowly cook the cherries until they are tender. Then start adding sugar, a little at a time, and tasting. When the fruit is sweet enough, cook for another 2 minutes. These cherries are delicious served warm with a little whipped cream on top.

Or, for something a little fancier, try this version of cherries jubilee. After sugaring and cooking the cherries, strain them and keep the juices. For each cup of cold juice, add another tablespoon of sugar and 1 tablespoon cornstarch. Make certain that the cornstarch is thoroughly dissolved before slowly heating, constantly stirring, until the mixture becomes clear and thickened. Then add 2 cups of the cooked cherries and heat through. Pour ½ cup of warmed brandy over the cherries and serve flaming with rich vanilla ice cream.

CHOKECHERRY
Prunus virginiana

The chokecherry gets its name from its v e r y sour taste, the strong acidity of which sometimes produces a constriction of the throat if the sourness catches you off guard. Although n o t as widely used by Indian tribes as some of the sweeter cherries, chokecherries were reputedly dried with their pits removed, p o u n d e d and leached, and used in the making of pemmican. The chokecherry has the same hard wood as the black rum cherry, but the trees are too small to attract lumbermen.

There a r e t w o places where the chokecherry really comes into its own. First, it is reported to be the most widespread species of tree in North America, growing from Alaska through Canada and the States into Mexico, and from one coast to the other. Second, it makes one of the tartest and most refreshing jellies ever tried on fresh corn bread.

The chokecherry is a large shrub or a small tree, seldom reaching over twenty feet in height. Its bark, when stripped, is not aromatic like that of the rum cherry. The dullish leaves are smooth, dark green above, and paler beneath. Oval to slightly oblong in shape, they have pointed tips, sharply pointed teeth on the margins, and are quite thin.

Clusters of delicate white flowers adorn the tree from early to late spring, depending on locality, when the leaves are almost fully expanded. In July to September, the flowers are replaced by clusters of deep red to reddish-purple and sometimes black fruits, each about the size of a pea. The clusters of cherries are usually shorter than the grape-like clusters of the rum cherry.

The chokecherry is an inhabitant of edges of woods and fields, in thickets, on stream banks, and along fencerows. It can often be collected in amazing abundance as the quantity of cherries on the branches make them bend precariously under the weight of the fruit.

For a delicious and colorful jelly, cook in separate saucepans, peeled, cored and quartered apples (preferably tart jelly apples), and mashed chokecherries, each with just enough water to cover them and keep them from burning. Pass each, when soft, through a jelly bag. For the jelly, one can use any proportions of these juices desired. We like them almost half-and-half, but perhaps for the first try it is better to use 1 cup chokecherry juice to 2 cups apple juice. To this combination of juices, add 3 cups of sugar and bring to a rapid boil. Boil until the sheeting stage is reached, then pour into glass sealers and seal with paraffin.

PIN-CHERRY

Prunus pensylvanica

Pin-cherries have a thin and quite acidic fleshy portion, with very large stones in comparison to the size of the fruit, hence they are often overlooked by persons out to collect the makings of a wild cherry pie. But their flavor and color are so fine in a jelly, that they are well worth seeking in their northern habitat.

A native of this continent, the pin-cherry also bears the names bird cherry (perhaps because of its appeal as a feed for many wild avian species) and fire cherry (perhaps again because it is quick to colonize

burned-over areas where light intensity is great). It is a lover of sunlight, and hence tends to situate itself in more open areas such as clearings, edges of fields and woods, roadsides, and fence-rows. The pin-cherry is quite common from Newfoundland and Labrador to British Columbia, and south at higher altitudes as far as North Carolina and Colorado.

The pin-cherry is similar in appearance to the others. It can be a tree reaching to thirty-five feet in height or a coarse shrub. Its thin bark, smooth when young but becoming progressively rougher with age, is a rich reddish-brown, and can easily be stripped from the tree. Its leaves are oval to oblong-shaped, with wide bases and pointed tips, thin, sharply toothed, shiny on top, and duller beneath. The white flowers appear from early spring to early summer, depending on locality, about the same time as the leaves, and grow in lateral, loose, somewhat flattened clusters. The flowers mature quite early to produce long-stemmed, light red, rounded fruit which varies in size from one specimen to the next.

Pin-cherries make a lovely, tart, brightly colored jelly. However, there is not much pectin in the fruit, so it is best to either mix the cherries with apples, or to add a package of pectin for every 4 cups of boiled juice. We prefer to add the pectin and enjoy the sharply sour taste of the "cherries only" jelly. To make the jelly, you can follow the same general directions given for chokecherry jelly.

If you have the time and inclination one afternoon, and want to make really great jelly doughnuts, try this recipe. First dissolve 1 package (or 1 level tablespoon) of dry active yeast in 1 cup of lukewarm milk and let it stand for 5 minutes. Then add 1 teaspoon salt, 2 cups flour, and beat until smooth. Cover the mixing bowl and let stand to rise for 30 minutes. Next, add ¼ cup vegetable oil, 2 well beaten eggs, 1 cup brown sugar, ½ cup powdered milk, and 1 more cup flour. Mix thoroughly, cover, and let rise again. When the dough is high and light, punch it down. Then turn it out onto a pastry sheet. The dough should be just stiff enough to handle. If it is too soft, mix in a little more flour. Divide the dough into two portions, and again let it rest for 10 minutes.

Now, roll the dough so that it is about ½ inch thick. With the rim of a highball glass (or with a doughnut cutter if you really want to be elegant), cut rounds out of the dough. For each pair of rounds, cover one with a heaping spoonful of pin-cherry jelly, sprinkle with a touch of cinnamon, and cover with the second round. Brush the edges of the rounds with a beaten egg, and press them together so that a seal is made to keep the jelly in place. When the doughnuts have all been fashioned and filled, place them on a piece of waxed paper, uncovered, and let them rise for 1 hour. Then fry them, 2 or 3 at a time, in a deep fryer filled with a light vegetable oil. When they are done, drain them well on paper towels and sprinkle with icing sugar.

PLUM

Prunus spp.

In eastern Europe, plums are greatly appreciated. They make excellent preserves, decorate sweetcakes like nothing else can, and produce a hearty and robust brandy that in places is hailed as the national drink. Here they don't seem to be as popular. Domestic plums in the supermarket are often overlooked by the family grocery shopper, and wild plums in the uplands are regularly bypassed by the wild harvester. Perhaps we need some re-education. Plums are not only nutritious and wholesome, but also extremely flavorful — especially the kind you collect yourself.

There are over a dozen species of wild plums in North America, all quite similar in form to their cousins, the wild cherries. Two of the better known of these are the beach plum, *Prunus maritima,* and the wild American plum, *Prunus americana.* Both are highly branched

shrubs, the former being quite short, while the latter sometimes takes the form of a tree over some twenty feet in height.

The wild American plum grows from the East Coast to Montana and Colorado, and south to the Gulf Coast. In the north, it can be found in southern Ontario and in the midwestern and Great Lakes states. It is a resident of thickets, fencerows, edges of woods, and stream banks. This small tree is somewhat thorny and quite shaggy of bark. The leaves are oblong, rounded at the bottom and pointed towards the tip, and sharply toothed. Masses of whitish spring flowers give rise to fruits that are close to an inch in length, juicy and pulpy, with a flattish stone. The fruit ranges from yellow to red in color, and is the predecessor of many of our cultivated plum varieties.

The beach plum generally does not have thorns. The short shrubs have saw-toothed, oval-shaped leaves, white flower clusters that normally appear before the leaves, and fruits that range from one-half to one inch in diameter. The fruits are juicy, pulpy, and a dull purple to blueish-purple in color, often with a bloom. One type bears yellowish fruit. The beach plum is an inhabitant of coastal sandy soils from New Brunswick to Delaware, and somewhat inland.

Westerners also have a wild plum, the Sierra or California plum (*Prunus subcordata*), which can be harvested in the south of Oregon and the very northern part of California.

Plums can be handled in a number of ways for the table. Some are good raw, but others are a bit acidic and have tough skins. All are good cooked or canned. One of the most popular ways to can wild plums is to wash and clean them, prick the skins, and pack them in glass sealers.

Cover the fruit with a hot light sugar syrup, seal, and process in boiling water for 20 minutes. Large plums can be halved and the stone removed before canning.

Stewed plums are delicious. To stew the fruit, halve and pit the plums, cover with just enough water to keep from burning, and simmer for about 20 minutes. Then add sugar to taste, a dash of cinnamon, and a touch of brandy. These can be eaten as a dessert, or as a side dish with moose, deer, or caribou.

For a dessert that always calls for seconds, butter a deep baking dish and line it with 1 inch of stewed or canned plums. In a separate bowl, mix ¼ cup flour, ¼ cup powdered milk, ¾ cup oatmeal, ½ teaspoon each of cinnamon and nutmeg, a dash of salt, and 1 cup of brown sugar (packed). Then, with a pastry blender, cut in ½ cup of vegetable shortening. Spread the crumb mixture over the plums, dot with butter, and bake in a moderate oven for ½ hour. Serve this hot, in deep dishes with cream poured over each serving.

Plums make good jams and jellies as well, either alone or in combination with other fruits.

GRAPE

Vitis spp.

Many wildlife species use the wild grapes for food, protection, and nesting cover. They are very hospitable plants, and are widely distributed across this continent. Some score or more species inhabit the temperate zones of Canada and abound through most of the United States wherever moist, fertile soil is found. Grapes are commonly seen growing along fences and stone walls, at the borders of woods and along stream banks, and on sandy, well-drained soil.

The wild grapes are too well known to warrant much of a description. They are mostly woody vines, climbing or trailing in habit by means of tendrils. Their simple leaves are broad, rounded to palmately lobed, and a bright green. The fragrant flowers mature to form clusters of blue to purple berries, pulpy but with numerous seeds. In short, the wild grapes greatly resemble their domestic counterparts, but generally have smaller leaves and smaller, yet still juicy, fruits.

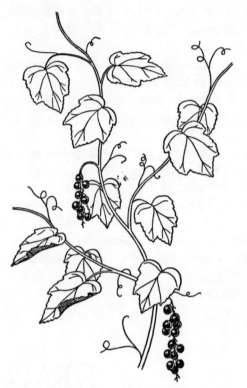

For jams and jellies, the northern fox grape, *Vitis labrusca,* is hard to beat. Its fruit is large and sweet, and grows in great abundance. In the south, the preference often turns to the southern fox grape or muscadine, *Vitis rotundifolia.* But no matter which species is used, the results are always good. The grapes should be picked over to ensure that no stems or odd bits of debris get into the preserves. Then they should be washed.

For a lightly spiced wild grape jelly, stew a couple of quarts of mashed grapes (without crushing the seeds) in as little water as possible for about 15 minutes. Then pass through a coarse strainer and a jelly bag. To each cup of grape juice, add ¾ to 1 cup sugar (depending on taste), ½ teaspoon each of cinnamon, allspice, and ground cloves, and liquid pectin (1 package per 4 cups). Boil rapidly for a couple of minutes, and then pour the jelly into sterilized sealers.

Wild grapes are very flavorful when combined with game. As a stuffing for quail or Hungarian partridge, they contribute towards the plump juiciness of the finished product. They are also good in sauces for sautéed game birds. We have long favored grouse baked while wrapped in grape leaves. And indeed we are also very fond of grape leaf rolls stuffed with rice and ground venison.

If you would prefer grapes for dessert, try a fresh grape pie. Starting with 5 cups of fresh grapes, remove the skins from the fruit by "popping" the flesh out into a saucepan. Drop the skins into the blender jar and buzz briefly to chop. Cook the mashed fruit for 3 minutes, then pass it through a sieve to remove the seeds. In a bowl, combine the fruit and the skins, along with 1½ tablespoons of finely grated lemon peel. To this, add a combination of 1½ cups sugar, ⅓ cup flour, and a dash of salt and cinnamon. Pour the grape mixture into an unbaked pastry shell. Dot with butter and cover with a second pastry crust. Bake the pie in a hot oven for about 45 minutes, until the crust is golden. Serve warm with ice cream or sour cream.

WINTERGREEN
Gaultheria spp.

While the long-famous oil of wintergreen was at one time extracted from wintergreen plants, this is no longer the case. Indeed, that sharp, tingly taste that you find with wintergreen toothpaste and chewing gum comes either from a distilled preparation of black birch twigs or is altogether synthetically manufactured. But the wintergreen still dresses up

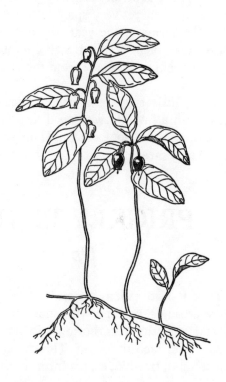

the winter countryside with its evergreen leaves and its bright red fruit, and is still a favorite among outdoorsmen as a winter snack.

Several species of wintergreen grow on this continent, but the aromatic wintergreen, *Gaultheria procumbens,* is perhaps the best known. This is an extensively creeping, usually herbaceous plant, with slender stems and erect flowering branches growing to a half foot or so in height. The aromatic, evergreen leaves, which are narrowly oval-shaped with small teeth on the margins, are quite soft and tender in spring, but become tougher as summer progresses. The leaves grow in clusters at the tops of the branches. From the leaf axils appear few white flowers, nodding on elongated stems. The red fruits are very interesting morphologically. Although they look like berries at first glance, they are actually capsules in which some of the flower parts have become fleshy and surrounded the capsule to give the appearance of a berry. The fruits are persistent on the plants throughout winter.

The aromatic wintergreen is a plant of rough, sterile country, including mountain clearings and evergreen woods. Its range stretches from Newfoundland to Manitoba, and south from New England to Minnesota, and into Georgia and Alabama. The plant has a host of common names, including partridgeberry, checkerberry, teaberry, boxberry, mountain tea, and at least a dozen others.

Several western species of wintergreen exist as well, perhaps the most commonly sought being *Gaultheria shallon* and *Gaultheria ovatifolia*. The latter species is similar to the aromatic one; the former much bigger. However, the leaves and berries of the western types are not as strongly flavored, although their berries are generally larger.

Wintergreen fruits have a somewhat sweet yet very spicy taste. They are quite refreshing when eaten in the spring of the year, right off the plants. In fact, their taste is really superior after the freezing winter. Some people prefer them in pies; others enjoy tea made from the young leaves.

We stew the berries briefly, add a little sugar and lemon juice to taste, and use them as a condiment with a favorite venison dish that is basted in mustard and cooked with onions and sour cream. Their spicy taste is also good with other heavy game meats.

PRICKLY PEAR

Opuntia spp.

The prickly pear, or Indian fig as it is also called, certainly does not look like a choice wild edible. It is a cactus, covered with short, irritating hairs and barbed bristles, and sometimes harboring solitary or grouped, sharply pointed spines up to an inch in length. The modified stems of the plant produce the typical, flat, tree-like cactus form so often

pictured in cartoons. The leaves are scale-like, and usually go unnoticed.

As with many of the cacti, the flowers of the prickly pear are beautiful. They are large, yellow, and generally quite short-lived. The pear-shaped fruits are generally tanned to reddish, and contain a fleshy, juicy pulp that is good to eat. However, the annoying little hairs and bristles also appear on the fruit, so if you are going out collecting, don't forget your plastic-faced gardening gloves and your knife.

The prickly pears are plants of arid soils, generally found growing in dry sands and rocky areas. They are common in the tropical regions of this continent, stretching their range to the temperate region in some areas. Their spread is from Florida to California, and some species are even reported as far north as Minnesota.

The prickly pears must be peeled before eating. They are best right out of hand, but some people stew them or cook them in other ways.

Chapter V

NATURE'S NUTTY SNACKS

Nature produces a wide range of nuts and seeds that make delicious out-of-hand snacks. They are, for the most part, ready to harvest at a delightful time of year, when the leaves of the forest are a blaze of crimson, orange, and yellow. The time of Indian summer, the smell of the autumn woods, the industriousness of wild creatures preparing for winter, and the crisp crackling of drying vegetation underfoot all add to a wild nut harvest.

The wild nuts are usually easily recognized. Many, like the hazelnuts, resemble the ones found in supermarkets. Others, such as the acorns, are old familiars. They generally require little special treatment after the harvest, and can be used in any recipe calling for nuts. Roasted and sealed in airtight containers, wild nuts and seeds will keep for a long time. They can also be frozen for added storage life. The wild nuts and seeds are full of flavor, and are among the most nutritious of the wild edibles on this continent.

SUNFLOWER

Helianthus spp.

The tall and stately sunflower, with its multitude of seeds in the center of the flower head, hardly needs to be described to most wild harvesters. The large, composite flower, actually made up of many ray flowers resembling petals around the outside and numerous disk florets bearing seeds on the inside, reaches three to six inches in diameter in the wilds. The seed-bearing disks range from one to two inches. The flower heads are large, but nothing like the size of the cultivated varieties.

There are numerous species of sunflowers in North America, all fairly close in resemblance. All are tall plants, growing up to six feet in height. The stalks are coarse and rough; the leaves coarse, toothed, and quite shaggy in appearance.

The seeds of the sunflower are quite nutritious, and are an important food source for many gallinaceous and migratory birds on this continent. Even some big game species are known to eat the plant. The common sunflower, the state flower of Kansas, is a native of meadows, plains, bottomlands, and other rich soil areas. Through escape from cultivation, the plant has spread to many roadsides, ditches, fencerows, a n d other waste places across the majority of the United States and southern Canada. The stronghold of the sunflowers, though, is in the western prairies.

Thin-leaf Sunflower

The Indians had many uses for sunflowers. The Hopi tribe used the plant to alleviate symptoms of spider bites. The Ojibwas used crushed sunflower roots to draw blisters. Many Indian tribes obtained dyes from the plant — black and purple from the seeds, and yellow from other parts — for ornamental use. Sunflowers have even been used medicinally as a diuretic, in combating the high fevers of malaria, and in easing bronchial problems.

As an economic plant, the sunflower has always been useful. The Iroquois and some western tribes beat the ripe seeds, boiled the mash in water, and thus separated sunflower seed oil by skimming the oil product from the surface of the water after cooling. This can still be done today, for those who are patient enough to carry out the whole process. In our generation, most prefer to collect their cooking oil from the grocer's shelf.

Roasted sunflower seeds were at one time a very popular snacking commodity on this continent. In some areas of Europe they still are. However, here they seem to have become somewhat unfashionable — no doubt due to the fact that roasting and shelling the little devils to get the meat out is a tedious job. But kids generally find that roasting and cracking sunflower seeds can be as much fun on a Sunday afternoon as popping corn.

If you are trying to cope with a mass of seeds at one time, it is easiest to put the seeds in a heavy plastic bag and break the shells with a hammer or a rolling pin. Then place the mash in a large container of water and stir vigorously until the nut meats sink to the bottom and the shells rise to float at the top. Now roast the nuts and sprinkle with a little salt. This handy snack can be eaten at any time or used in any recipe calling for nut meats.

If you have used the latter "quick" method to shell your sunflower seeds, don't throw the shells away. Try them in sunflower seed coffee. The Seneca Indians used roasted sunflower seed hulls, once the meats had been removed, to make a coffee-like beverage, simply by pouring hot water over the hulls. Early explorers quickly picked up the idea, and claimed the brew to taste "just like coffee." We have tried it, and found the explorers to be not far wrong.

Lastly, the sunflower can be regarded as one of the best emergency foods available. The seeds need not be roasted; they can be eaten raw. And young flower heads, before the seeds are fully formed, provide a good source of protein when simply boiled and eaten like broccoli.

ACORN
Quercus spp.

The oaks are without a doubt among the most beautiful and stately trees that grow on this continent. A great number of species and interspecies hybrids exist, in a great range of habitats and soil types, and at many different elevations. The fruits of these trees are quite popular with a variety of wildlife species, including upland game birds, squirrels and chipmunks, bears, sheep, and deer.

Among many tribes of Indians, the fruits were at one time one of the most important nut foods. Yet acorns don't seem to be terribly popular as table fare today. Perhaps this is because most people don't know where to begin with them in the kitchen.

All of the acorns are edible, yet understandably some are better than others. Some are sweet, yet quite a number are bitter due to a high tannin content. Before the acorns can be successfully used as a nut food, some of the tannins must be leached out. To do this, shell the acorns and put them into a large saucepan. Cover them with water and boil. Change the water as it becomes yellowed with the tannins. After the tannins have been removed, drain the acorns and place them on a heavy cookie sheet. Then dry them very slowly in the oven, mixing them often to ensure that they don't become scorched. Once they have dried, they can be stored and then chopped or ground and used in any recipe calling for nuts.

White Oak

The easiest way to recognize an oak tree is when it has acorns on it. Almost everyone knows what an acorn looks like. Most people also recognize the typical "oak" leaf. But not all of the oaks have such leaves. Some are distinctly different.

There are basically two groups of oaks in North America — the white oaks and the black or red oaks. The fruits of the former are generally sweeter and make better eating, those of the latter nearly always being bitter. The nuts of the white oaks are also smooth inside, while those of the black or red oaks are downy or furry. The fruits of the white oaks take one year to produce; those of the blacks or reds take two.

The trunk bark of the white oaks is usually scaly and off-greyish in color. That of the black oaks is typically much darker, harder, and distinctly furrowed. The leaf tips and lobes of the white oaks are often rounded and generally without bristly points; those of the blacks have bristly elongations. Of course, not all oak leaves are lobed; some are oblong and even lance-shaped.

Although when one thinks of the mighty oak, one thinks of a very tall and full tree, it must be remembered that some oaks are shrubs that may never grow over a dozen feet in height. All of the oaks are hardwoods, and some of the larger species have long been prized and sought by lumbermen.

As I mentioned earlier, dried acorns can be used in any dish calling for nuts. One of our favorites, for breakfast and for lunch-box fare, is acorn and currant muffins. For these, combine 2 cups cake and pastry flour with 4 teaspoons baking powder, ½ teaspoon salt, and ¼ cup sugar. Add ½ cup chopped acorns and ½ cup dried wild currants. In a separate bowl, mix 2 eggs, ¼ cup melted butter, and 1 cup milk. Stir the wet ingredients into the dry, just until barely combined. The batter should be lumpy, and over-stirring will make for tough muffins. Pour the batter into greased muffin tins or paper muffin cups. Twenty minutes in a hot oven and your muffins will be ready.

HAZELNUT
Corylus spp.

The hazelnuts, or filberts as the cultivated type are often called in the supermarkets, are highly branched shrubs or small trees, at times reaching upwards of twenty feet in height. The twigs and branches are popular with moose and deer, while the nuts are the favorites of squirrels and chipmunks. Hazel is common in our part of the country, and is often found in prime grouse habitat. In fact, we have shot many a grouse whose crop has been filled with the long spikes (called catkins) of this native plant.

The catkins of the hazels appear in early spring, before the foliage unfolds. The alternate leaves are oval-shaped and wide, with toothed margins and sharply pointed tips. The nuts, which bear a leafy husk, ripen in late summer and early fall.

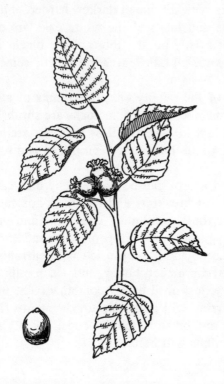

The two most common species of hazelnuts are the American hazelnut (*Corylus americana*), which is primarily an eastern species, and the beaked hazelnut (*Corylus cornuta*), which grows from east to west. In the fruit of the American hazel, the husk is open and flares at the top. The nuts have brown shells that are usually hard and thick. On the other hand, the fruit of the beaked hazel, as the name suggests, has a long and bristly husk that contracts into a long neck or "beak." The shell of the nut is considerably lighter in color, sometimes almost a whitish-brown, thin, and easily cracked.

The American hazelnut is a plant of thickets and edges of woods, growing from Maine to Saskatchewan, and south as far as Florida and Oklahoma. The beaked hazelnut ranges a bit farther north, from Newfoundland to British Columbia, and south to Georgia, Kansas, and Oregon. It is a plant of thickets, borders of woods, fencerows, and clearings. A variety of the beaked hazelnut, sometimes considered a separate species, also grows in the mountains of California.

One of the most elegant and delicious cakes to serve after a wild goose dinner is a wild hazelnut torte. To start, separate 6 eggs. Beat the yolks until they are thick and light in color. Then, still beating, gradually add ½ cup of sugar, followed by 1 cup of ground hazelnuts and ⅓ cup of fine dry bread crumbs. Beat well.

In a separate bowl, beat the egg whites with ¼ cup sugar until they form stiff peaks. Gently fold the yolks and whites together, starting with ¼ of the whites only, until well combined. Then pour the batter into a greased and floured spring-form pan, and bake it at 325° F for about 40 minutes or until done. Remove from the oven to a cake rack and cool completely. When the cake has cooled, divide it into three horizontally with a long piece of thread.

In the meantime, beat 2 cups of heavy whipping cream with ¼ cup sugar, 1 cup ground hazelnuts, and ½ teaspoon rum flavoring. Use this filling between the layers and around the edges of the cooled cake. For the top of the cake, beat 2 eggs with 4 tablespoons of sugar until thick and light. Beat in ⅓ cup of soft butter and 2 squares of melted, semi-sweet chocolate. Spread the smooth frosting over the top of the torte, cover with slivers of hazelnuts, and refrigerate for at least ½ hour before serving.

BLACK WALNUT
Juglans spp.

There are several species of walnuts in North America. For the most part, they are tall and majestic trees, although some of the western species are large shrubs. The eastern black walnut, *Juglans nigra*, is one of the best known and most widely spread. Its range extends from western New York through southern Ontario to Minnesota, and south to Texas and Florida.

The black walnut is a tall tree of deep, rich soils, with strong, hard w o o d o f straight grain that is favored both by cabinetmakers and gunstock craftsmen. Its bark becomes d e e p l y browned with age, and the stout twigs are somewhat hairy. The alternate leaves are made up of fifteen to twenty-three leaflets borne in pairs. Each leaflet is oblong with a sharp tip and toothed margins. Often the terminal leaflet does not exist, making the number of leaflets even.

The fruits of the black walnut are hard to miss. And once you have handled them, you will be easy to spot, too. The husks of the walnuts contain a lasting dark brown dye that stains everything it touches and leaves telltale marks on your hands for over a week. This dye, however, as I discovered during a stint of working in a histology laboratory, is removable with biological stain solvents available from most chemical supply houses. This is really handy to know if you have been out collecting walnuts on a Saturday morning and have a heavy date that night!

The black walnuts grow up to two or more inches in diameter, and are enclosed in a thick, yellowish-green husk that makes the whole fruit rounded in shape. The nuts grow singly, or in clusters of two and seldom three. At maturity, they fall from the tree. The walnut itself resembles the furrowed, four-celled nut that is commonly tossed into fudge and brownies, and cracked in front of the open fire while watching the Rose Bowl game.

The green husks of the walnuts can be removed with a knife, or the husks can be left to dry before cracking. If the former method of dehusking is chosen, rubber gloves are a wise precaution against semipermanently tanned hands.

The wild walnuts can be eaten fresh or baked, exactly as the store-bought kind. One treat that has become a favorite in our house is made by creaming ½ cup of butter with 1 cup of dark brown sugar. Then beat in ¼ teaspoon salt, 1 teaspoon vanilla, and 2 egg yolks. Next, mix in 1¼ cups flour that has previously been sifted with 2 teaspoons baking powder. When this mixture is smooth, pour it into a shallow, greased baking pan.

In another bowl, beat the 2 egg whites until stiff. Then beat in ¾ cup light brown sugar, and stir in 1¼ cups chopped walnuts, and ½ teaspoon of vanilla. Spread this over the first layer and bake in a moderate oven for about 30 minutes or until done. This cake must be cooled before cutting.

BUTTERNUT

Juglans cinerea

The butternut is a very close relative of the walnuts, differing from the walnuts mainly in that the husked fruits are elongated rather than rounded, and the thin husks are sticky and covered with hairs. The butternut, sometimes called the white walnut, is not nearly as impressive a tree as the black walnut, nor is its wood as good. The wood is somewhat more coarsely grained, softer, weaker, and much lighter in color.

The bark of the butternut is greyish, smooth, and becomes broadly furrowed with age. The leaves resemble those of the black walnut, except with eleven to seventeen leaflets, and the odd one usually present. The sharply ridged, oblong nut, which is sometimes over two inches in length, contains a sweet, oily kernel, hence the local common name of oilnut. The husked fruits are borne in clusters of two to five and become brown and drop at maturity.

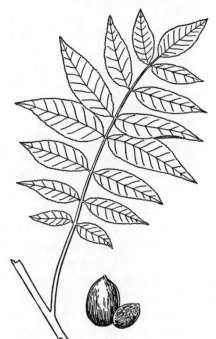

The butternut is an inhabitant of rich, well-drailed soil areas. Its range extends from southwestern New Brunswick through Ontario to North Dakota, and south to Georgia and Arkansas.

For a really rich, nutty tidbit to serve with coffee and brandy after a roasted fowl dinner, combine 2 cups of sugar, 1 cup of cereal cream, and 1 heaping tablespoon of butter in a saucepan and cook to the soft ball candy stage. Then add ½ pound of pitted dates, very finely chopped, and a touch of salt. Cook over very low heat for another 15 minutes, stirring constantly, then stir in 2½ cups chopped butternuts. Cool, and shape the confection into a long, thin roll, wrapped in waxed paper. Chill in the refrigerator, and then slice thinly before serving.

HICKORY
Carya spp.

The hickories belong to the same botanical family as the walnuts, but differ from their relatives in that the husks enclosing the nuts crack open when mature, resulting in four partitions that reveal the smooth, ripened nut. Upwards of a dozen hickories grow on this continent, and all are edible. Some, of course, are better than others.

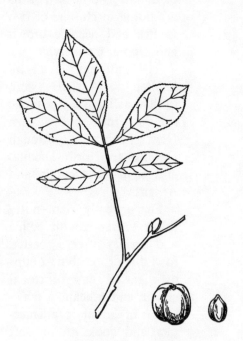

Shagbark Hickory

The familiar pecan of the supermarkets is one of the hickories. It grows wild in the bottomlands of the lower Mississippi River valley, producing a thin-shelled nut the shape of an olive. This species, *Carya illinoensis,* is a moderately tall tree.

Of the sweet hickories, the shagbark or shellbark, *Carya ovata,* is probably the best known. This t r e e reaches over one hundred feet at times, and is easily recognized by its bark. The bark of young stems is smooth and pale grey, while that of older trunks separates into long, shaggy strips that come loose at one or both ends. The wood of the tree is strong, hard, and finely grained, and makes an excellent choice for a durable and reliable ax handle. Wood chips of this and the other sweet hickories are also excellent in the smoke house, producing delicious hams, fish, and fowl.

The alternate leaves of the shagbark hickory are compound, usually consisting of five leaflets (though sometimes three or seven), borne in opposite pairs. The lance-shaped leaflets are sharply pointed and finely toothed. The fruit matures in mid- to late autumn. The rounded, thick husk is a deep reddish-brown, while the thin-shelled nut is whitish and very sweet.

The shagbark hickory is usually found in moist and fertile yet well-drained soil. It commonly inhabits hillsides and valleys, borders of swamps, and rich, mixed woods. Its range extends from Maine, through southern Quebec and Ontario, to Minnesota and Nebraska, and south to Florida and Texas.

There are the bitter hickories as well. As the name indicates, these sometimes have bitter nuts, but often they are very good to eat. The bitter hickories, or pignuts as they are sometimes called, have nuts with thin husks. These husks generally crack only at the top, but sometimes crack fully late in the season. The wood of these trees is also hard and strong, and these species are found over much the same range as the sweet hickories.

Because of their thin, easily cracked shells and deliciously sweet meat, hickory nuts are very easily handled with a nutcracker or a small hammer. And for a real Sunday dinner treat, try a hickory nut pie. Cream ⅓ cup butter, gradually beating in ¾ cup brown sugar (packed) until light and fluffy. Then gradually beat in 3 eggs. Finally, stir in 1 cup light corn syrup, 1 cup chopped hickory nuts, 1 teaspoon vanilla, and a dash of salt. Pour this mixture into an unbaked pie shell, and bake in a moderately hot oven for about 35 minutes. Cool sufficiently before serving.

BEECH

Fagus grandifolia

The beechnuts were one of the most important nut foods among the Indians and the early settlers alike. They were used fresh and dried, and even roasted for coffee by the colonists. At one time their range spread into a good portion of the Midwest, but today the one native species of beech that exists on this continent is pretty well relegated to the east. Its range stretches from the Canadian Maritime Provinces to the southern end of Lake Superior, and south below the Mason-Dixon line. One variety grows as far south as Florida and Texas.

The American or red beech, more commonly simply called the beech, is a dweller of rich bottomlands and upland areas, as well as well-drained soil areas such as hillsides and ridges. It is rarely found in swamps.

The tree is an attractive one, quite tall and with a smooth, light bluish-grey, dappled bark. Its hard, strong wood has often been used for railway ties and barrel staves. The alternate leaves occur singly. Oblong to oval in shape, they have pointed tips and margins with sharply incurved teeth. The nuts ripen in October. They are triangular, shiny brown, sharply pointed, and usually borne in pairs in a bristly brown husk that splits open at maturity. Beechnuts are thin shelled and are reported to contain over twenty percent protein. Their edibility is confirmed by the genus name *Fagus,* which is derived from the Greek word *phagein,* meaning to eat.

Everyone likes brownies, and they are made even better with wild beechnuts added. Melt 2 squares (2 ounces) of unsweetened chocolate with ¼ cup of butter. Remove from the heat and stir in 1 cup sugar, 2 unbeaten eggs, a dash of salt, ½ cup flour, 1 teaspoon vanilla, ¾ cup chopped beechnuts, and ¼ cup dried currants. Spread this mixture into a greased, non-stick baking pan, and bake in a moderate oven for about ½ hour. Be careful not to overbake. Cool and cut into squares.

Chapter VI

PLANTS TO STAY AWAY FROM

The average wild harvester, if he or she displays just a little caution and common sense, should not have any problems with the poisonous plants. Most are easily recognized by key characteristics of their anatomy, the habitat in which they live, or the way in which they grow. At any rate, there are not all that many species of poisonous plants in North America that are likely to be mistaken for wild edibles.

The "poisonous" plants can be divided broadly into two categories — those that cause illness when eaten, and those that cause inflammation or discomfort if touched. Examples of the latter are poison ivy, poison oak, and poison sumac, and are amply described in any number of books on the outdoors. Anyone venturing into the uplands in search of wild edibles should become familiar with these species.

The former species are those that have been known to cause illness or even death if eaten. Wild harvesters should come to know these as well, and a number of them are described briefly in this chapter.

One thing that must be remembered about the so-called poisonous plants is the old cliché about one man's poison being another man's food. Different persons display varying degrees of susceptibility to different toxic agents, and, of course, quantities have a great deal to do with the degree of illness that can be produced by consuming them.

In addition, it should be remembered that some wild edibles like the marsh marigold require cooking before they can safely be eaten. Others, like the May apple or the pokeweed, have only certain parts that are edible, the remainder of the plant being poisonous. These types of plants have been specifically mentioned in the previous chapters.

Generally speaking, the plants described in *this* chapter should be shied away from. They are the most common of the poisonous plants, and the wild harvester should not take chances with them.

WATER HEMLOCK. *(Cicuta maculata).* The water hemlock is one of the most troublesome poisonous plants in North America because its roots have an odor much like parsnips, and children, among others, often mistake them for young wild parsnips. Perhaps it should be stated at this point that children should be allowed to partake in the wild harvest only under the guidance of an adult or someone knowledgeable in edible wild plants. The water hemlock is a plant of swales, moist meadows, and shallow water areas. It somewhat resembles the wild carrot, but has a thicker and taller stem and much coarser leaves. Growing up to eight feet in height, the lower of its alternate leaves are often tri-forked with toothed leaflets. The white flower clusters are shaped like flat-topped umbrellas. The fruits are small, dry, plump, and have grooved surfaces. The roots often consist of a group of fleshy, tuber-like branches. A bulb-bearing water hemlock also exists in swamps and shallow water areas. The upper branches of this plant bear small, bulbous projections.

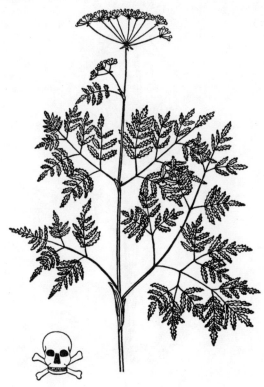

POISON HEMLOCK. *(Conium maculatum).* The foliage of the poison hemlock resembles that of parsley. The plant, however, contains a toxic alkaloid material which makes it poisonous. A dweller of waste areas, including garbage dumps and old quarries, the poison hemlock is densely spotted with sticky purple blotches of varying sizes. The stem is coarse and round; the leaves finely divided like those of the carrot. White flowers grow in flat-topped umbrella-shaped clusters.

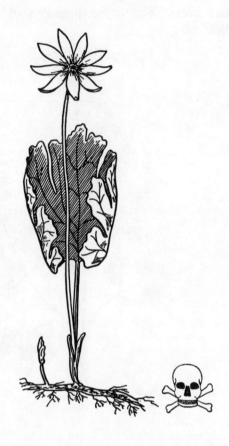

BLOODROOT. *(Sanguinaria canadensis).* The bloodroot is one of the first harbingers of spring. Its deeply lobed leaves and single-flowered stems, which arise from a perennial rhizome, are known to almost all who visit the spring woods. But no matter how interesting the rootstock looks, it is poisonous and should be left alone.

WHITE HELLEBORE. *(Veratrum viride).* As was mentioned in the discussion of the marsh marigold, poisonous alkaloids render the white hellebore inedible. Also known as Indian poke, this plant grows in low-lying woods, moist meadows, and along stream and river banks. Its large, overlapping leaves are alternate and somewhat conspicuously veined. The quick-growing plants produce long, loosely branched clusters of greenish flowers. The white hellebore grows in the same habitat as skunk cabbage, marsh marigold, and watercress.

DOGBANE. (*Apocynum* spp.). There are a number of dogbanes in North America, all with milky sap. For this reason, they are at times confused with the milkweeds. However, one of their distinguishing features is their rapidly forking stem with opposite, oblong to oval-shaped leaves arising directly from the stem. The plant bears loose, spreading clusters of bell-shaped flowers. An inhabitant of stream banks and edges of woods, the dogbanes should be completely avoided.

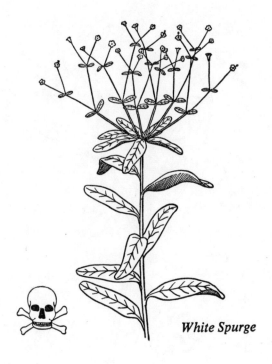

White Spurge

SPURGE. (*Euphorbia* spp.). The spurges are an odd lot of plants. Mostly herbs of the temperate and tropical regions, the many species are often strikingly dissimilar in appearance, and range from small bushes to trees. Yet they all have the common link of a milky, acrid juice, and hence are sometimes mistaken for the milkweeds. Just how poisonous the spurges are is not really known. The Christmas poinsettia, which is a member of the spurges and was long thought to be poisonous, has recently been proven harmless. But the wild spurges should be kept away from.

JIMSON WEED. *(Datura stramonium).* Another member of the night-shade family, the foul-smelling jimson weed has been widely publicized as a hallucinogen. The plant is a coarse annual, loosely branched and sometimes reaching four feet in height. Its large, toothed leaves and petunia-like flowers are often seen in cosmopolitan areas such as vacant lots, railroad yards, and other waste places. The somewhat prickly egg-shaped capsules contain lentil-like seeds, and have a papery sheath at the base. The seeds are particularly poisonous, containing two alkaloid toxins.

The foregoing are but a few plants in North America that are known to have deleterious effects when eaten. Most of the poisonous plants, indeed, display themselves quite readily with either a foul odor or a bad taste. So it is unlikely that any adult would pursue them as a wild edible. Anyone venturing into the fields and forests to collect edible plants should be cautious of species with which they are not familiar. This is one of the cases where an ounce of prevention is worth more than a ton of cure.

NIGHTSHADE. (*Solanum* spp.). You have no doubt heard of the deadly nightshade. This red-berried plant is a woody climber, common in thickets and clearings near civilization. Its appearance varies widely, and even includes a form resembling the potato. The nightshades contain a poisonous glucoside that render the plants inedible.

GLOSSARY OF TERMS

ALTERNATE (leaves). Borne not opposite to one another, yet at regular intervals successively on opposite sides of the stem.

ANNUAL. A plant that flowers, produces seed, and dies in its first year.

BIENNIAL. A plant that grows vegetatively in its first year, overwinters, and flowers, produces seed, and dies in its second year.

BRACT. A specialized or modified leaf occurring beneath a flower or flower cluster.

CATKIN. A dry, scaly flower spike, usually of one sex.

COMPOSITE (flower). An inflorescence or head composed of many tiny flowers, usually of different types. A member of the family *Compositae*.

FLOWERING SPIKE. A series of simple flowers growing directly from an elongated stem or spike.

FROND. The expanded leaf-like part of a fern.

HERBACEOUS. Herb-like. Not woody.

INFLORESCENCE. The flowering portion of a plant.

LEAF AXIL. The angle formed between the stem of a plant and the leaf or leaf stalk.

OPPOSITE (leaves). Occurring opposite to one another on the stem.

PERENNIAL. Flowering year after year before dying.

PETIOLE. The stalk attaching a leaf to the stem.

RHIZOME. A prostrate or underground stem portion which gives rise to roots and shoots at intervals.

ROSETTE. A circular cluster of leaves, usually low to the ground.

SHRUB. A woody perennial plant, usually smaller than a tree and with numerous main upright branches or stems.

185

STIPULE. A small appendage at the base of a leaf or a leaf stalk.

STOLON. An above-ground basal branch or runner that roots at intervals.

TAPROOT. The thick, primary descending root which gives rise to smaller roots.

TENDRIL. A slender, clasping or twining growth, usually used to cling and climb.

TREE. A woody perennial plant, with one main upright stem, namely a trunk.

TUBER. An underground branch, usually thick and short, and having eyes or buds.

VINE. A climbing woody perennial, usually attaching itself by means of tendrils.

WINTER ANNUAL. A plant that is sown in autumn, overwinters, and flowers and produces seed the following spring.

INDEX TO RECIPES

INDEX TO PLANT NAMES

189

Two Other Important Books from Pagurian Press

JEROME KNAP
THE COMPLETE OUTDOORSMAN'S HANDBOOK
A Guide to Outdoor Living and Wilderness Survival

Everything you need to know for outdoor survival: interpreting animal actions, recognizing poisonous plants and wildlife hazards, understanding a compass and map reading, as well as the skills of canoeing, archery, and snowshoeing.

Jerome Knap is one of the country's foremost outdoor writers. He is also the author of *The Hunter's Handbook, Training the Versatile Gun Dog,* and *Getting Hooked on Fishing.*

JEROME KNAP
WHERE TO FISH AND HUNT IN NORTH AMERICA
Including Mexico and the Caribbean

The Complete Sportsman's Guide

This complete guide is devoted to providing the latest, and most up-to-date information on fishing, hunting, and where to go for best results. Wherever you are or want to go across the United States and Canada, first check out your itinerary with this invaluable guide: what licences are needed, when is open season, what local laws or restrictions must you know, who is a reliable guide, how much will it all cost?